S0-CDQ-225

ALLE ZEIT WACH
1842

Peter Nijkamp · Aura Reggiani

Interaction, Evolution and Chaos in Space

With 60 Figures

Springer-Verlag
Berlin Heidelberg New York
London Paris Tokyo
Hong Kong Barcelona
Budapest

Professor Dr. PETER NIJKAMP
Faculty of Economics and Econometrics
Free University
De Boelelaan 1105
NL-1081 HV Amsterdam, The Netherlands

Dr. AURA REGGIANI
Faculty of Economics
University of Bergamo
Via Salvecchio, 19
I-24100 Bergamo, Italy

ISBN 3-540-55458-0 Springer-Verlag Berlin Heidelberg New York Tokyo
ISBN 0-387-55458-0 Springer-Verlag New York Berlin Heidelberg Tokyo

Library of Congress Cataloging-in-Publication Data. Nijkamp, Peter. Interaction, evolution, and chaos in space / Peter Nijkamp, Aura Reggiani. p.cm. Includes bibliographical references and index.
ISBN 3-540-55458-0 (Berlin : alk. paper). -- ISBN 0-387-55458-0 (New York : alk. paper) 1. Spatial analysis (Statistics) 2. Chaotic behavior in systems. 3. Econometrics. I. Reggiani, Aura. II. Title.
QA278.2.N54 1992 330'.01'5118--dc20 92-13515 CIP.

Printing: Weihert-Druck, Darmstadt
Bookbinding: J. Schäffer GmbH u. Co. KG., Grünstadt
2142/7130-543210 - Printed on acid - free paper

This book is dedicated to those whose spatial interaction is governed by strange attractors

PREFACE

For many decades scholars from various disciplines have been intrigued by the question whether there are unifying principles or models that have a validity in different disciplines. The building of such analytical frameworks bridging the gaps between scientific traditions is a very ambitious task and has not been very successful up till now.

In the past - in a static context - several such principles have been defined and advocated at the edge of the natural sciences on the one hand and social sciences (in particular, economics and geography) on the other hand, mainly based on the paradigm of 'social physics'. Some important contributions to the integration of the spatial systems sciences and physics can be found in gravity theory and entropy theory, which have formed the corner stones of interaction models in space.

This book is about spatial interaction models. It describes the origin, the history and the correspondence of such models from a 'social physics' perspective. It is emphasized that such models need a clear behavioural underpinning as a sine qua non for a valid use in spatial systems analysis. This view also explains the use of micro-based disaggregate choice models as a tool for analyzing spatial systems. This is mainly analyzed in Part A of this book.

In recent years much attention has been devoted to qualitative (structural) changes in dynamic systems, evolutionary theory, morphogenesis, bio-social sciences and the like. Also here the question emerges as to the validity of such approaches in social sciences in general and in spatial systems in particular. Is it, for instance, possible to design models that describe indigenous behavioural shocks in spatial systems models?

In this context, non-linear dynamics has many important lessons to offer to the analysis of the dynamic behaviour of spatial systems. Especially modern chaos theory, which has gained much popularity in recent years, presents fascinating new analytical departures. At the same time the need for a behavioural explanation in such qualitative structural change models has to be emphasized. Therefore, this book contains in the second part (Part B) a critical overview of chaos types of models, with a particular emphasis on the applicability in dynamic spatial systems.

We conclude in our book that spatial interaction models - interpreted in a broad sense - may offer a general framework for many (static and dynamic) phenomena in interconnected spatial systems. They appear to be compatible with 'social physics' and chaos principles. A major challenge is now to generate a more solid empirical basis for such models.

On our 'chaotic' way to the final product in this book we have been guided by many colleagues and friends. Furthermore, our thanks go to Dianne Biederberg, Rita Hittema and Pilar van Breda who managed to create order out of the seemingly chaotic pieces which formed the building blocks of the present study and had to be typed in a logical and consistent order. We also wish to recognize the support offered by the Netherlands Institute for Advanced Study in the Humanities and Social Sciences (NIAS) in Wassenaar and the Italian Consiglio Nazionale delle Ricerche (CNR) (no. 90.02147.CT11 and no. 91.02288.CT11), while completing this book.

Wassenaar, February 1992

Peter Nijkamp
Aura Reggiani

CONTENTS

PART A STATIC MODELS OF SPATIAL INTERACTION

1. SPATIAL INTERACTION MODELS AND GRAVITY THEORY A CONCISE OVERVIEW
1.1 Introduction 3
1.2 Gravity Analysis and Spatial Interaction Models 3
1.3 Gravity Theory and the Social Sciences 8
1.4 Alternative Utility Foundations and Specifications of Gravity Theory 10
1.4.1 Simple interaction theory 10
1.4.2 System-wide cost efficiency 12
1.4.3 Aggregate utility theory 13
1.5 The Scope of Gravity Models: Concluding Remarks 16

2. ENTROPY THEORY AND SPATIAL INTERACTION ANALYSIS
2.1 Prologue 17
2.2 Entropy Theory and Spatial Interaction 17
2.3 Alternative Specifications of the Entropy Model 25
2.4 Alternative Theoretical Backgrounds of the Entropy Model 30
2.4.1 An economic utility approach 30
2.4.2 A probabilistic utility approach 31
2.4.3 Statistical information theory 34
2.4.4 Bayesian statistics 35
2.4.5 Maximum likelihood approach 37
2.5 Concluding Remarks 38

3. ENTROPY AND GENERALIZED COST MINIMIZATION MODELS AT THE MACRO LEVEL
3.1 Prologue 39
3.2 Entropy and Linear Programming 40
3.3 Entropy and Geometric Programming 42
3.4 Spatial Patterns of Entropy and Linear Programming Models 47
3.5 Entropy Revisited 52
3.6 Concluding Remarks 54

Annex 3A. Relationships Between Total Trip Costs and the Cost Friction Coefficient 56

4. **SPATIAL INTERACTION MODELS AND UTILITY MAXIMIZING BEHAVIOUR AT THE MICRO LEVEL**
4.1 Prologue 59
4.2 Spatial Interaction Behaviour and Individual Choice Behaviour: Theory 59
4.2.1 Introduction 59
4.2.2 Spatial interaction models and deterministic utility theory 62
4.2.3 Spatial interaction models and random utility theory 63
4.2.3.1 Basic concepts of random utility theory 63
4.2.3.2 Analogies between spatial interaction models and discrete choice models 68
4.2.4 Concluding remarks 71
4.3 Spatial Interaction Behaviour and Individual Choice Theory: An Application 72
4.3.1 Introduction 72
4.3.2 The model 72
4.3.3 The data 75
4.3.4 Results and concluding remarks 75
4.4 Conclusions 82

Annex 4A. An Algorithm for Modal Split Choice with Congestion 84

PART B DYNAMIC MODELS OF SPATIAL INTERACTION

5. **DYNAMIC AND STOCHASTIC SPATIAL INTERACTION MODELS**
5.1 Prologue 89
5.2 Spatial Interaction Models Analyzed by Means of Optimal Control 90
5.2.1 Introduction 90
5.2.2 An optimal control approach 91
5.2.3 Concluding remarks 94
5.3 Spatial Interaction Models Analyzed by Means of Stochastic Optimal Control 95
5.3.1 Introduction 95
5.3.2 A stochastic optimal control approach 97
5.3.3 Concluding remarks 102
5.4 Spatial Interaction Models with Catastrophe Behaviour Analyzed in the Framework of Stochastic Optimal Control 102
5.4.1 The model 102
5.4.2 The stochastic optimal control version 104
5.5 Epilogue 107

Annex 5A. The Generalized Spatial Interaction Model as a Solution to the Optimal Control Entropy Model 109

Annex 5B. A (Generalized) Stochastic Spatial Interaction Model as a Solution to a Stochastic Optimal Control Problem 113

Annex 5C. Stability and Bifurcations in a Phase Diagram Analysis for a Stochastic Optimal Control Problem 116

6. SPATIAL MODELLING AND CHAOS THEORY
6.1 Prologue 119
6.2 Chaos Theory: A Brief Review 120
6.2.1 A general introduction to non-linear modelling 120
6.2.2 Key issues in the theory of chaos 125
6.3 Spatial Applications of Chaos Theory: A Brief Survey 133
6.3.1 Introduction 133
6.3.2 Dendrinos 135
6.3.3 Dendrinos and Sonis 137
6.3.4 Mosekilde, Aracil and Allen 137
6.3.5 Nijkamp 138
6.3.6 Reiner, Munz, Haag and Weidlich 139
6.3.7 White 139
6.3.8 Zhang 140
6.3.9 Concluding remarks 140
6.4 A Model of Chaos for Spatial Interaction and Urban Dynamics 140
6.4.1 Introduction 140
6.4.2 Results of simulation experiments 143
6.4.2.1 The onset of chaotic motion 143
6.4.2.2 Chaotic urban evolution 147
6.4.3 Concluding remarks 149
6.5 Epilogue 150

Annex 6A. Classification of Two-dimensional Critical Points 152

Annex 6B. Strange Attractors: A Brief Overview 155

Annex 6C. Steady State Solutions for a Generalized Lorenz System 161

7. SPATIAL INTERACTION MODELS AND CHAOS THEORY
7.1 Prologue 165
7.2 Chaos in Spatial Interaction Models 166
7.2.1 Introduction 166
7.2.2 Chaotic elements in dynamic logit model: theory 166
7.2.3 Simulation experiments for a dynamic logit model 170
7.2.3.1 Dynamic processes in logit models 171
7.2.3.2 Dynamic processes in spatial interaction models 174
7.2.4 Concluding remarks 179
7.3 Delay Effects in Dynamic (Binary) Logit Models 180
7.3.1 Introduction 180
7.3.2 A logistic model with multiple delays 181
7.3.3 Concluding remarks 191
7.4 Conclusions 192

Annex 7A. Stability Solutions for a Dynamic Logit Model 193

Annex 7B. Stability Solutions for a Dynamic Spatial Interaction Model 197

8. **SPATIAL INTERACTION ANALYSIS AND ECOLOGICALLY-BASED MODELS**

8.1 Prologue 199
8.2 Prey-Predator Models: Introduction 200
8.3 Synergetic Models of Spatial Interaction 203
8.4 An Optimal Control Model for a Spatial Prey-Predator System 206
8.4.1 Introduction 206
8.4.2 Equilibrium analysis 207
8.4.3 Concluding remarks 209
8.5 Competition Models: Introduction 210
8.6 Impact of Chaotic Evolution in Spatial Competition 214
8.6.1 Introduction 214
8.6.2 The case of two competing regions 214
8.6.2.1 Equilibrium analysis 214
8.6.2.2 Simulation experiments 217
8.6.3 Concluding remarks 221
8.7 Epilogue 222

Annex 8A. Stability Solutions for an Optimal Control Prey-Predator Problem 223

Annex 8B. Transformation of a Continuous System into a Discrete System 228

Annex 8C. Stability Analysis for a Particular Competing System 230

9. **RETROSPECT AND PROSPECT**

9.1 Retrospect 233
9.2 Typology of Dynamic Spatial Interaction Models 237
9.2.1 Introduction 237
9.2.2 Macro-dynamic approaches 238
9.2.3 Micro-meso dynamic approaches 240
9.3 Evolution of Spatial Interaction Models 241
9.4 New Research Areas 243

References 245

Index 275

PART A

STATIC MODELS OF SPATIAL INTERACTION

CHAPTER 1

SPATIAL INTERACTION MODELS AND GRAVITY THEORY
A CONCISE OVERVIEW

1.1 Introduction

Economic geography and regional economics have since their inception focused attention on the analysis of spatial patterns of human activities, in both the production and the household field. This major field of scientific research is predominantly mirrored in location theory.

In recent decades however, the focus of interest has shifted towards the analysis of spatial movements, i.e. the processes or spatial flows emerging as a result of given spatial configurations. Especially the popularity of Spatial Interaction Models (SIMs) reflects the importance attached to spatial processes in analyzing spatial systems (see, for instance, Masser and Brown, 1977). SIMs are essentially models of spatial flows, i.e. flows of people, goods, information, from some origin i to some destination j (cf. Sen, 1985). Such models have become the most important analytical tools for studying the 'geography of movement' (see Lowe and Moryades, 1975), and they are nowadays a prominent descriptive device for investigating the dispersion and coherence of activities in a spatial system (see Horowitz, 1985).

SIMs have various roots; even a conventional linear programming model or a spatial input-output model may be regarded as a specific SIM, while also network equilibrium models (e.g., combined models of location, destination, mode and route choice) may be interpreted as SIMs in a broader context (see Boyce et al., 1983). But the most common specification of a SIM originates from gravity theory or from entropy theory. Both theories will briefly be described in Chapter 1 and Chapter 2, respectively.

1.2 Gravity Analysis and Spatial Interaction Models

Gravity theory can, from a methodological viewpoint, be considered as a relational theory (cf. Harvey, 1969), which describes the degree of spatial interaction between two or more points in space in a manner analogous to physical phenomena. A central element in classical gravity theory is formed by Newton's law from physics, which states that the attraction force a_{ij} between two entities i and j is proportional to their respective masses m_i and m_j and inversely proportional to the squared distance, d_{ij}^2, between these entities. Consequently, this law of gravitational attraction states:

$$a_{ij} = \gamma\, m_i\, m_j\, d_{ij}^{-2} \tag{1.1}$$

where γ is a constant proportionality factor.

One of the first foundation stones for the use of gravity theory in the social sciences can be found in Carey (1858), who stated: "Man tends, of necessity, to gravitate towards his fellow man" (p.42). The first attempt to apply some concepts from Newtonian physics to spatial interaction flows can be found in Ravenstein (1885) in a study on migration flows between English cities. In the first part of this century interesting applications of the gravity concept were developed by Young (1924) and Reilly (1931) in the area of farm population migration and trade flows in the retail sector, respectively. First complete specifications of spatial analogies of Newton's law were presented by Stewart (1941) and Zipf (1949). Their models had the following general form:

$$T_{ij} = \gamma\, O_i\, D_j\, d_{ij}^{-2} \tag{1.2}$$

where T_{ij} is the degree of spatial interaction from i to j (e.g., migration flows, telephone calls, trade flows, passenger traffic etc.). The proportionality constant γ may be regarded as a scalar which adjusts the model to the units of measurement of the (known) variables O_i and D_j, which stand for stock variables (e.g., population size) in the places of origin and destination, respectively. It is noteworthy that, apart from dimensionality reasons, the constant γ was interpreted by Stewart as the total energy of interaction of i with respect to j.

From the above early applications of gravity theory to spatial interaction analysis, it is thus clear that a straightforward analogy to Newtonian physics is assumed (see Foot, 1986). This analogy was once more justified, as empirical applications in many fields supported statistically a gravity formulation for spatial flows. A rigorous legitimation of gravity theory in spatial analysis was given by Isard at the First Regional Science Association Meeting in 1954 (see Isard and Maclaren, 1982). Since the 1950s much theoretical and empirical work has been pursued in the area of SIMs and of formulations emerging from gravity theory (see, for a broad review, Fotheringham and O'Kelly, 1989, Haynes and Fotheringham, 1984, and De La Barra, 1990).

It is clear that the impact of the spatial discount factor in (1.1) or (1.2) is not necessarily equal to the exponent 2 in the distance friction function. Furthermore, it may also be plausible to assume a more general Cobb-Douglas specification for the impact of interacting stocks i and j. This may lead to the following more general specification of a gravity model for spatial interactions:

$$T_{ij} = \gamma O_i^{\alpha_i} D_j^{\alpha_j} d_{ij}^{-\beta} \tag{1.3}$$

It is assumed in (1.3) that the flows T_{ij} are calculated as a weighted geometric average of O_i and D_j, in contrast to an unweighted geometric average in (1.3). This observation is important, as the inequality relationship between a geometric average and a normal arithmetic average is one of the critical points of departure for **geometric** programming (cf. Hardy et al., 1967) which forms the basis for a **dual** optimizing interpretation of SIMs (see Chapter 3).

In general terms, a SIM can be represented as follows:

$$T_{ij} = f_i(O_i)\, f_j(D_j)\, F_{ij} \tag{1.4}$$

where $f_i(O_i)$ and $f_j(D_j)$ are compound origin and destination factors, respectively, and where F_{ij} is the generalized distance friction effect between i and j. This general formulation subsumes not only the conventional gravity model, but also related specifications such as the well-known **intervening opportunities** model (see Stouffer, 1949, 1969, and for a review Zipser, 1990). If $f_i(O_i)$ and $f_j(D_j)$ are regarded as constants (see, for instance, equations (1.2) or (1.3)), inconsistencies will arise, because then the additivity conditions (i.e., the sum total of individual flows equals the total volume of flows) are violated (see Somermeyer, 1966). This means that (1.4) has to be reformulated in order to meet the above mentioned additivity conditions, which imply in case of a gravity specification for a SIM:

$$\sum_j T_{ij} = O_i \qquad \forall i \tag{1.5}$$

and

$$\sum_i T_{ij} = D_j \qquad \forall j \tag{1.6}$$

Now it is easily seen that the additivity conditions are satisfied if the gravity model is specified as follows:

$$T_{ij} = A_i\, B_j\, f_i(O_i)\, f_j(D_j)\, F_{ij} \tag{1.7}$$

where γ is now replaced by the two balancing factors A_i and B_j, which are equal to:

$$A_i = \{ \sum_j B_j\, f_j(D_j)\, F_{ij}\}^{-1}\, O_i\, \{f_i(O_i)\}^{-1} \tag{1.8}$$

and

$$B_j = \{ \sum_i A_i\, f_i(O_i)\, F_{ij}\}^{-1}\, D_j\, \{f_j(D_j)\}^{-1} \tag{1.9}$$

Clearly, without such balancing factors, a doubling of both all O_i's and D_j's would quadruple the volume of trips between all points i and j, an almost absurd forecast caused by the multiplicative structure of a gravity model (see Gould, 1972). Next, it is evident that in case of the standard gravity model:

$$T_{ij} = A_i\, B_j\, O_i\, D_j\, F_{ij} \tag{1.10}$$

the balancing factors A_i and B_j are equal to:

$$A_i = (\sum_j B_j\, D_j\, F_{ij})^{-1} \tag{1.11}$$

and

$$B_j = (\sum_i A_i\, O_i\, F_{ij})^{-1} \tag{1.12}$$

It should be added that in the conventional gravity models for spatial interaction analysis, the distance friction function F_{ij} has usually the following form:

$$F_{ij} = \exp(-\beta\, c_{ij}) \tag{1.13}$$

where c_{ij} stands for the unit distance friction costs between point i and j (e.g., travel costs, travel time, psychological costs etc). It should be noted here that specifications different from the exponential function expressed in (1.13) have also been applied in the literature (for example, Gamma functions, Tanner functions, power functions; see, for a review, Choukroun, 1975, and Erlander and Stewart, 1990). The criterion for the choice of the deterrence function is essentially a pragmatic one (see Openshaw, 1977). However, it is interesting to underline here that the form of the deterrence function is strongly influenced by the spatial (geographical) configuration of the opportunities leading to the interaction. In this context the exponential structure seems well suitable for describing a homogeneous-isotropic system (without strong barriers or metropolitan areas) (see Diappi and Reggiani, 1985).

Especially specification (1.10) (and the related constraints (1.11) and (1.12)) has become a common analytical tool in spatial interaction analysis (e.g., migration, transportation). It should be admitted however, that also some alternative interesting SIMs not belonging to the standard gravity family have been developed, such as the competing destinations model (see Fotheringham, 1983, and Gordon, 1985). Nevertheless, gravity theory has undoubtedly formed the basis of spatial interaction analysis (see also Erlander, 1980).

In empirical applications the unknown balancing factors A_i and B_j can be estimated by means of recursive solution techniques (see e.g., Batty, 1970a, 1970b, Batty and Mackie, 1973, Bouchard and Pyers, 1964, Chisholm and O'Sullivan, 1973, and Wilson et al., 1969). The basis for this method was already provided by Deming and Stephan (1940) and Furness (1965). Convergence conditions of this method have been studied amongst others by Evans and Kirby (1974), MacGill (1977, 1979) and Sen (1982). The balancing factors of a conventional (doubly constrained) SIM are often interpreted as a measure of accessibility (in terms of attractiveness and repulsion) of one zone with respect to other zones (see Kirby, 1970).

The estimation of the distance friction coefficient, β, can be pursued by using either least squares methods (see, for instance, Cesario, 1975, Flowerdew and Aitken, 1982, Gray and Sen, 1983, Sen and Pruthi, 1983, and Sen and Soot, 1981) or maximum likelihood methods (see for instance Batty, 1976, Evans, 1971, Evans and Kirby, 1974, and Hyman, 1969).

Finally, it should be noted that particular cases of model (1.10) are the singly-constrained SIM and the unconstrained SIM. The first model (a so-called production-constrained SIM) can be represented by the following equation:

$$T_{ij} = A_i \, O_i \, D_j \, F_{ij} \tag{1.14}$$

where the balancing factor A_i now has the following specification:

$$A_i = (\sum_j D_j \, F_{ij})^{-1} \tag{1.15}$$

The value of A_i serves to ensure that model (1.14) reproduces the volume of flows originating from zone i, so that:

$$O_i = \sum_j T_{ij} \tag{1.16}$$

Therefore, model (1.14) is an allocation-interaction model, since the fixed activity O_i is allocated to destination zones, whilst at the same time also the interaction matrix T_{ij} is produced.

Analogously to the production-constrained SIM we also have an attraction-constrained SIM where the destination zone totals are fixed and the model is used to determine the level of activity generated from the origin zones. The formulation of this type of model is therefore straightforward.

The last formulation of a SIM is the unconstrained type, in which neither the origin zone totals nor the destination zone totals are fixed. It can be essentially represented by equation (1.4) where f_i (O_i) and f_j (D_j) are regarded as constants.

1.3 Gravity Theory and the Social Sciences

Newtonian physics forms the basis for the application of gravity theory in spatial interaction analysis. The popularity of this so-called 'social physics' poses also important questions of a methodological nature, such as the theoretical justification of this approach, the 'mechanistic' basis of gravity theory, and the simplification of social reality by means of physically-based gravity models. A simplification of social reality by means of physically-based gravity models. A justification of the use of social physics requires the fulfilment of various methodological conditions (see also Crilly, 1974, and Heggie, 1969):

(1) The method should satisfy normal **logical conditions** for the social system at hand (for example, spatial additivity conditions and non-negativity for the various flows).

(2) There should be a high degree of **plausible correspondence** between the phenomenon described in the physical world and that from a socio-spatial system.

(3) The physical basis of the approach should be interpreted in terms of **social aspects** of the system at hand (for example, in terms of spatial actions, social behavioural hypotheses, or at least in terms of statistical regularities of aggregate spatial choices of the social system at hand).

(4) The mathematical specification of the various relationships should correspond to **plausible hypotheses** about the behaviour of spatial (inter)actors.

(5) The empirical results of a physically-based spatial interaction approach should be **confirmed by data from social reality**.

Condition (1) is not a priori guaranteed by the traditional gravity model. However, as shown in (1.8) - (1.9) the introduction of balancing factors ensures the fulfilment of the additivity conditions in a spatial interaction system, which at the same time ensure non-negative values of all variables.

Condition (2) is normally fulfilled by the traditional gravity model: there is a strong formal analogy, if not a direct correspondence, between the movements of particles in a physical system and the point-to-point movements of units in a spatial interaction system; in both systems the behaviour of elements is determined by push and pull factors in addition to the friction caused by distance.

Condition (3) is more difficult to meet, as there is a priori no logical reason why the behaviour and motives of people should be similar to those of particles in physics: choice behaviour is an essential ingredient of social sciences and this can hardly be represented by means of a physical theory. Nevertheless, it may be possible to find a formal behavioural rationale for the use and specification of a gravity model (see Section 1.4).

Condition (4) is closely linked to the third condition and will be re-examined in the next section.

Condition (5) appears to be fulfilled in many empirical applications. The gravity model appears to be a rather robust mirror of observed spatial configurations. It is reasonable to state that the popularity of gravity models is mainly due to its good empirical results (apart from its simplicity) and that the search for a theoretical foundation and justification has started only after numerous satisfactory outcomes in practice.

In conclusion, gravity theory appears to fulfil conditions (1), (2) and (5). The fulfilment of conditions (3) and (4) is somewhat more doubtful. Therefore, in the next section an attempt will be made to examine whether it is possible to derive a gravity theory in which conditions (3) and (4) are also satisfied.

The conclusion from the above observations is that social physics may be a valid and plausible paradigm in spatial interaction analysis. This is confirmed by an introductory statement of Fotheringham and O'Kelly (1989) who claim that spatial interaction is "movement or communication over space that results from a decision process....... In each case an individual trades off in some manner the benefit of the interaction with the costs that are necessary in overcoming the spatial separation between the individual and his/her possible destination" (p.1).

This 'behavioural' basis of SIMs explains also the wide spectrum of applications of such models, for instance, in the area of land use analysis, residential choice theory, migration analysis, housing market research, industrial location analysis, transportation planning, labour market research, retail market analysis, etc.

1.4 Alternative Utility Foundations and Specifications of Gravity Theory

It is evident that there is a need to firmly root the use of 'social physics' in behavioural social science theory. In addition to the considerations given in Section 1.3, various alternative ways have been developed to give a methodological justification for the use of gravity theory in the social sciences. Such attempts will briefly be described here.

1.4.1 Simple interaction theory

It is evident that any utility foundation of gravity theory should be based on push and pull effects as well as on distance deterrence effects between points of origin and destination in space.

Starting from scratch, the gravity model can be derived by abandoning any physical analogy or physical basis and by turning instead to plausible hypotheses about spatial interactions (see for example Isard, 1960). Assume a country with I cities i (i=1,....,I) and with a homogeneous population. The total population is P, and the population size of city i is denoted by P_i. The total number of interactions (e.g., trips) between all cities i and j is assumed to be equal to T. Furthermore, for the moment we assume all movements to be costless.

Given these assumptions, the trips made by an individual from city i to j will depend on the size of i and j. It is clear that for this individual the expected proportion of trips to city j is equal to P_j / P. Furthermore, it is clear that the expected (average) number of trips made by any arbitrary individual is equal to T/P (due to the homogeneity assumption). This number can be set equal to a constant k. Therefore, the expected number of trips per individual from any city i to a given city j is equal to kP_j / P. Hence, without a distance deterrence function, the total number of expected trips, T^*_{ij}, from city i to j is:

$$T^*_{ij} = k\, P_i\, P_j \,/\, P \tag{1.17}$$

Next, the assumptions of costless movement and of a homogeneous population may be relaxed, by assuming that the ratio of actual to expected trips (T_{ij} / T^*_{ij}) is an inverse function of distance:

$$T_{ij} \,/\, T^*_{ij} = c\, d_{ij}^{-\beta} \tag{1.18}$$

Substitution of (1.17) into (1.18) gives:

$$T_{ij} = \gamma\, P_i\, P_j\, d_{ij}^{-\beta} \tag{1.19}$$

where γ is equal to:

$$\gamma = c\, T/P^2 \tag{1.20}$$

Therefore, the gravity model may also be derived on the basis of rather simple principles for spatial interaction. It should be noted, however, that the underlying assumptions (for example, equal trip rates independent of distance) are rather rigid and contain some implicit behavioural hypotheses about spatial behaviour which should be made more explicit (see Section 1.4.3).

1.4.2 System-wide cost efficiency

An alternative foundation for a theory of spatial interaction behaviour based on a gravity-type approach was provided by Smith (1978a, 1978b). By means of advanced topological notions he proposes a macro-statistical principle of spatial interaction, the so-called **cost efficiency principle**. This principle asserts in essence that, other things being equal, spatial flow patterns involving higher total costs are less likely to be observed than those involving lower costs. This principle appears to characterise completely the class of gravity models.

Given the cost efficiency principle as a macro-device for a spatial system, Smith even shows that from among all possible probabilistic models of trip behaviour which vary continuously with the underlying structure of travel costs, the only models which are consistent with the cost efficiency principle are the exponential gravity models.

The proof of this statement runs via several steps, including inter alia assumptions of a continuous probability distribution of trips as a function of travel costs, of trip independencies (i.e., absence of multi-purpose trips), of constant average travel costs, and of an inverse relationship between the trip pattern probabilities and the total travel costs of each trip pattern. Via a whole set of successive lemmas Smith shows ultimately that a trip distribution is cost efficient if and only if there exists an exponential gravity representation for this trip (cf. also Stetzer and Phipps, 1977).

The relevance of this approach lies in the fact that exponential gravity models are derived from macro-statistical principles which are more directly interpretable from a behavioural viewpoint. Smith's approach appears to satisfy our methodological conditions (1) (logic and consistency), (2) (high degree of formal analogy), (3) (behavioural interpretation), (4) (plausible mathematical specification) and (5) (link to reality). It should be noted, however, that the cost efficiency principle is only a macro-statistical principle. This principle does not imply that **individuals** tend to make less costly trips rather than more costly trips. In other words, micro utility-maximizing behaviour may be a foundation of the cost efficiency principle, but it has not been proven that micro-behaving units tend to minimize travel expenditures at each level of trip activity.

A second issue arises from the generality of the cost efficiency principle. While it states that less costly trip configurations are more probable than more costly trip configurations at a given level of trip activity, it does not specify **in which way** at a micro or macro level different distance factors and push and pull effects are evaluated by the spatial (inter)actors. In other words, the mathematical specification of a utility model for evaluating alternative trip costs is missing. One possible specification of such a utility model will now be discussed in the next subsection.

1.4.3 Aggregate utility theory

The question arises whether a behavioural criterion can be specified which can be interpreted as a utility function associated with a gravity model. Earlier contributions to an analysis of relationships between utility theory and gravity theory are contained among others in Golob and Beckmann (1971), Golob et al. (1973), Gordon (1985), Niedercorn and Bechdolt (1969), and Nijkamp (1975).

Suppose that each point of origin i can be regarded as an aggregate decision unit for trip decisions in a spatial interaction framework. Consequently, the hypothesis may be made that this decision unit has an aggregate welfare function (utility function or cost function), which may be regarded as a representative average welfare function for trip behaviour of the individuals living in this point of origin. The spatial decision unit is next assumed to allocate a certain aggregate 'travel budget' among a series of trips to alternative destinations. This travel budget is assumed to be a function of the number of individuals living at point i. Thus an optimal trip allocation is achieved by maximizing the above mentioned welfare function (which can be indicated as w_i) subject to the conditions arising from the travel budget.

Consequently, the utility-maximizing problem of point of origin i is:

$$\begin{aligned} &\max\ w_i = f_i\,(T_{i1},\ldots,T_{ij}) \\ &\text{subject to} \\ &\qquad \sum_j c_{ij}\,T_{ij} = C_i \end{aligned} \tag{1.21}$$

where c_{ij} represents the unit transportation cost between i and j and C_i the total travel budget of all individuals at point i. Next, C_i is assumed to be related to the total number of trip-makers at point i by means of a power function:

$$C_i = \eta\, O_i^{\alpha_i} \tag{1.22}$$

which implies that the travel budget is assumed to be proportional (via the parameter η) to the total push effect ($O_i^{\alpha_i}$) of place i. Substitution of (1.22) into (1.21) and application of a first-order maximizing procedure gives:

$$\begin{aligned} w'_{ij} &= \frac{\partial w_i}{\partial T_{ij}} \qquad \forall\, i,j \\ &= \lambda\, c_{ij} \end{aligned} \tag{1.23}$$

where λ is the Lagrange multiplier related to the budget condition. Now for all unknown variables from point i the following system of equations can be constructed:

$$\begin{aligned} w'_{i1} / c_{i1} &= w'_{iJ} / c_{iJ} \\ &\vdots \\ w'_{i,J-1} / c_{i,J-1} &= w'_{iJ} / c_{iJ} \\ \sum_j c_{ij}\, T_{ij} &= \eta\, O_i^{\alpha_i} \end{aligned} \tag{1.24}$$

If it were now possible to cast the gravity model (1.3) in such a form that it satisfies (1.24), then we would have a behavioural interpretation of the gravity model. It is easily seen that for point i, formulation (1.3) can be rewritten as follows:

$$\begin{aligned} \eta\, O_i^{\alpha_i} &= \eta\, \gamma^{-1}\, T_{ij}\, D_j^{-\alpha_j}\, d_{ij}^{\beta} \\ &= \sum_j c_{ij}\, T_{ij} \end{aligned} \tag{1.25}$$

Next, it can be directly checked that a sufficient condition for the fulfilment of relationship (1.25) is:

$$\eta\gamma^{-1}\, T_{ij}\, D_j^{-\alpha_j}\, d_{ij}^{\beta}\, \{D_1^{\alpha_1} (\sum_j D_j^{\alpha_j})^{-1}\} = c_{i1}\, T_{i1}$$

$$\vdots \qquad\qquad \vdots$$

$$\eta\gamma^{-1}\, T_{ij}\, D_j^{-\alpha_j}\, d_{ij}^{\beta}\, \{D_J^{\alpha_J} (\sum_j D_j^{\alpha_j})^{-1}\} = c_{iJ}\, T_{iJ} \qquad (1.26)$$

By dividing the first J-1 equations by the last equation one obtains:

$$D_1^{\alpha_1} / c_{i1}\, T_{i1} = D_j^{\alpha_J} / c_{iJ}\, T_{iJ}$$

$$\vdots \qquad\qquad \vdots \qquad (1.27)$$

$$D_{J-1}^{\alpha_{J-1}} / c_{i,J-1}\, T_{i,J-1} = D_j^{\alpha_J} / c_{iJ}\, T_{iJ}$$

Now it is evident that (1.27) can be regarded as a specification of the first-order maximizing conditions (1.23), if the welfare function w_i of point i has the following shape:

$$w_i = D_1^{\alpha_1} \ln T_{i1} + \ldots + D_J^{\alpha_J} \ln T_{iJ} \qquad (1.28)$$

This log-linear welfare function of the aggregate trip decisions satisfies also the second-order maximizing conditions. This welfare function includes the elements $D_j^{\alpha_j}$ ($\forall$j) as preference elasticities with respect to trip decisions. In other words, the pull effect of a point of destination j, measured by means of $D_j^{\alpha_j}$, is implicitly used as a parameter representing the preference weight of the aggregate trip utility function of point i. This result can be regarded as a reasonable and plausible utility interpretation of the gravity model.

Consequently, one may conclude that the use of a gravity model can be justified on the basis of an appropriate utility theory for aggregate trip behaviour. The implicit utility function appears to be a log-linear utility function with the pull effects of each destination point acting as the preference elasticities (although it should be noted that this function is not necessarily the only one which satisfies the first-order maximizing conditions for utility maximization due to the fact that only sufficient conditions were specified). Further discussions on the derivation of gravity-type trip distribution models can be found in Choukroun (1975). More recent contributions can be found in Leonardi (1981c) dealing with accessibility and congestion in spatial interaction systems and Roy and Brotchie (1984) dealing with combined supply-demand elements in such systems.

In conclusion, conventional Newtonian physics has opened a rich field of research in spatial interaction theory. Especially the emphasis on a behavioural theory of spatial interaction has led to path finding new departures for SIMs.

1.5 The Scope of Gravity Models: Concluding Remarks

From the 1960s onwards the use of social physics in urban, regional and transportation planning models became very fashionable and a wide range of applications in these areas can be found. Especially at the beginning a close link was established with the popular Lowry type of models (see Lowry, 1964). Such models were mainly developed for land use planning but found a fruitful synergy with gravity type of models. The Lowry model is essentially based on two singly-constrained SIMs (see, for a good review of the Lowry model and its subsequent developments Batty, 1972).

Early examples of such models developed in the USA and based on applications of the gravity concept to land use activity allocations are inter alia: TOMM (Time-Oriented Metropolitan Model) developed for Pittsburgh (Crecine, 1964), BASS (Bay Area Simulation Study) reported by Goldner and Graybeal (1965), PLUM (Projective Land Use Model) reported by Goldner (1968), and DRAM (Disaggregated Residential Allocation Model) for Philadelphia (Putman, 1977). In Europe operational work along these lines was carried out at Reading University (Batty, 1970a), at Cambridge University (Echenique et al., 1969), at the University of Leeds (Bonsall et al., 1977) and at the Centre for Environmental Studies in London (Barras et al., 1971). Some applications were also developed for Latin American cities, such as Santiago (Echenique and Domeyko, 1970) and Caracas (Echenique et al., 1974). These earlier studies were followed by an avalanche of literature in the 1970s and 1980s.

Despite the popularity of the gravity concept in regional science and related disciplines, the need for a more rigorous foundation of this concept remained. The theoretical basis of SIMs based on gravity theory was still regarded insufficiently strong. A further theoretical reflection on gravity analysis was stimulated by the pioneering work of Wilson (1967, 1970), who showed that SIMs can be derived from a mathematical optimization problem, in particular by maximizing an entropy function. This original approach will be discussed in Chapter 2.

CHAPTER 2
ENTROPY THEORY AND SPATIAL INTERACTION ANALYSIS

2.1 Prologue

In Chapter 1 the need for a behavioural, social science based underpinning of spatial interaction models has been outlined. It turned out that in the history of these models several attempts have been made to offer a plausible description of behavioural backgrounds implicitly present in these models. Such behavioural notions appeared to be particularly interesting from a macro-systems viewpoint. Individual motives (or utility functions) played a much less important role. Since the 1970's two main streams in spatial interaction research can be distinguished, the first one more macro-oriented and based on the entropy concept and the second one more micro-oriented and based on discrete choice models. In the present chapter the nature and use of the entropy concept will be described. This framework will be extended, in Chapter 3, to a utility interpretation offered by optimization models. Next, Chapter 4 will be devoted to a further exploration of the relationships between discrete choice models and spatial interaction analysis.

More specifically in this chapter on entropy, the history and backgrounds of this new 'social physics' in spatial interaction analysis will first concisely be described. Then the question of its social science background will be dealt with. Finally, some alternative approaches related to entropy models will be presented.

2.2 Entropy Theory and Spatial Interaction

The entropy concept originates essentially from thermo-dynamics (see Fast, 1970), and describes the most probable configuration of particles in a closed physical system. Such a system can in principle adopt numerous states. In general, however, the elements of such a system tend towards an arrangement which can be organized in as many ways as possible. Such a tendency towards a seemingly maximum 'disorder' (or more appropriately, variety) means essentially that for a closed physical system a situation of disorder is more probable than an ordered situation.

Thus entropy has in principle a probabilistic background, related to the statistical distribution of events in an uncertain situation. As a consequence, entropy has also played a role in other disciplines, such as ecology (see Odum, 1971) and information theory (see, amongst others, Hobson, 1971, Jaynes, 1957, Mathai and Rathie, 1975, and Theil, 1967). Seen from a probabilistic perspective, entropy in the context of information theory represents expected information; it mirrors the degree of uncertainty about the realization of events in information systems, where this uncertainty is indicated by means of a discrete probability distribution. A reduction of entropy means a reduction in uncertainty, so that additional information reduces entropy. The mathematical specification of entropy in information sciences bears a close resemblance to that used in physics, since both concepts originate from a common statistical source.

Since flows in spatial systems may also adopt a huge variety of configurations, it is conceivable that the entropy concept has also a relevance in analyzing spatial interactions. In the meantime the fields of applications are numerous: trip distribution, trade flows, migration flows, modal choice, activity allocations, consumer expenditures and so forth. The basic idea is that the spatial distribution of units or parts in a system may exhibit a great many arrangements, from which the statistically most probable configuration can be identified by using the principle of maximum entropy. Consequently, entropy in spatial interaction analysis is a probability concept describing the outcome of a stochastic spatial allocation process.

In accordance with the distinction set out above between entropy in physics and information theory, two uses of the entropy concept in regional analysis may be distinguished. First, entropy can be employed as a descriptive device, based on the assumption of a spatial equilibrium for an allocation process. A most probable state of a system then corresponds to a maximum entropy of the system. In this way one arrives at the general specification of a gravity model between spatial entities; see, for example, Wilson (1970), Batty (1970a), or Cordey-Hayes and Wilson (1971). Entropy-maximizing models permit the determination of the "most probable" spatial flows of commodities, people or information. By assessing the parameters associated with the resulting gravity model, one could use the entropy approach for the projection of future spatial flows. A second use of the concept of entropy is for measuring the degree of organization in a spatial system. The concept of entropy is then employed as a tool for studying spatial differentiation; for instance, by investigating whether certain spatial configurations are completely arbitrary and disordered, or whether these configurations show a certain degree of spatial organization or regularity. Early analyses in this area have been undertaken by Berry (1964), Medvedkov (1967), Gurevitch (1969) and Semple and Gauthier (1972).

In the context of spatial interaction analysis we will focus our attention in particular on the first use, by providing a macro-analytical description of the most probable state of individual units in a spatial allocation system. The theoretical basis for this approach was laid by Wilson (1967, 1970), who showed that the most likely macro distribution for micro spatial movements in a spatial system can be obtained by means of the entropy principle. The arguments run as follows.

Assume a spatial system composed of origin and destination points with uncertain flows T_{ij} for which the following standard additivity conditions hold (see also Chapter 1):

$$\sum_j T_{ij} = O_i \tag{2.1}$$

and

$$\sum_i T_{ij} = D_j \tag{2.2}$$

Besides, it is clear that the following consistency condition should also hold:

$$\sum_i \sum_j T_{ij} = \sum_i O_i = \sum_j D_j = T \tag{2.3}$$

Furthermore, the assumption is made that the spatial system concerned has an upper limit on its total transport cost budget, caused by physical distance, transport time, psychological perception costs etc. Without the assumption of distance frictions, any spatial system would have an infinite solution. Assuming again fixed unit transport costs c_{ij} between i and j, the travel budget reads as follows:

$$\sum_i \sum_j c_{ij} T_{ij} = C \tag{2.4}$$

Given the constraints (2.1)-(2.4), the question is now: which spatial distribution of trips is the most probable arrangement of the system concerned?

According to the entropy principle, the most probable arrangement is formed by the spatial configuration which possesses the greatest number of micro states associated with it. The number of ways in which individual units can be assigned to a particular origin-destination table is then the following combinatorial formulation:

$$\omega(T_{ij}) = \frac{T!}{\prod_i \prod_j T_{ij}!} \tag{2.5}$$

The maximum value of $\omega(T_{ij})$ represents the maximum number of assignments of individual units to an origin-destination matrix. This maximum number of states of a spatial configuration dominates the number of states of alternative arrangements to such a degree that the spatial allocation of the trips associated with this maximum is the most likely one. This confirms that the entropy approach is a probability approach: the "optimal" spatial arrangement of a system is formed by the most probable arrangement, which is characterized by the fact that the value of $\omega(T_{ij})$ in this arrangement is at a maximum.

Since $\omega(T_{ij})$ is invariant with a monotonically increasing transformation, the objective function of the spatial system can be written in a logarithmic form as follows:

$$\ln \omega(T_{ij}) = \ln T\,! - \sum_i \sum_j \ln T_{ij}\,! \tag{2.6}$$

or following Stirling's approximation:

$$\ln \omega(T_{ij}) = \ln T\,! - \sum_i \sum_j (T_{ij} \ln T_{ij} - T_{ij}) \tag{2.7}$$

Since ln T! is a constant, it need not be considered in the maximization procedure, so that the ultimate problem to be solved is as follows:

$$\max \ln \omega(T_{ij}) = - \sum_i \sum_j (T_{ij} \ln T_{ij} - T_{ij}) \tag{2.8}$$

subject to:

$$\begin{aligned} \sum_j T_{ij} &= O_i \\ \sum_i T_{ij} &= D_j \\ \sum_i \sum_j c_{ij} T_{ij} &= C \end{aligned} \tag{2.9}$$

It should be noted that the argument in formulation (2.8) is defined in statistical mechanics as the entropy H of the system. Since in our case $H = \ln \omega$ is the log of the probability of a trip distribution, entropy is monotically related to probability defined in this way (see also Wilson, 1967). Consequently, (2.8) can be regarded as a measure of **interactivity**, related to the freedom of choice of the trip makers. Thus a high level of interactivity means that the trips are scattered over many zones, while a low level of interactivity means that the trips are more centred (see, e.g., also Erlander and Stewart, 1990). Consequently, entropy can also be related to uncertainty, in the sense that it is related to the most probable distribution with the greatest number of microstates giving rise to it. The solution of problem (2.8) - (2.9) can now easily be obtained by constructing a Lagrangian function L for this constrained maximization problem, i.e.,

$$\begin{aligned} L = \ln \omega(T_{ij}) + \sum_i \lambda_i (O_i - \sum_j T_{ij}) + \sum_j \mu_j (D_j - \sum_i T_{ij}) \\ + \beta (C - \sum_i \sum_j c_{ij} T_{ij}) \end{aligned} \tag{2.10}$$

where $\lambda_{i,}$ μ_j and β are the Lagrange multipliers associated with (2.1), (2.2) and (2.4), respectively.

The necessary conditions for the maximum are as follows:

$$\frac{\partial L}{\partial T_{ij}} = - \ln T_{ij} - \lambda_i - \mu_j - \beta c_{ij} = 0 \tag{2.11}$$

or

$$T_{ij} = \exp(- \lambda_i - \mu_j - \beta c_{ij}) \tag{2.12}$$

Next, by writing

$$A_i = \exp(- \lambda_i) / O_i \tag{2.13}$$

and

$$B_j = \exp(- \mu_j) / D_j \tag{2.14}$$

one obtains

$$T_{ij} = A_i B_j O_i D_j \exp(- \beta c_{ij}) \tag{2.15}$$

By making use of expressions (2.1) and (2.2) it is easily seen that the balancing factors A_i and B_j are equal to:

$$A_i = \{ \sum_j B_j D_j \exp(- \beta c_{ij})\}^{-1} \tag{2.16}$$

and

$$B_j = \{ \sum_i A_i O_i \exp(- \beta c_{ij})\}^{-1} \tag{2.17}$$

It can be proven that the second-order conditions for a maximum are also satisfied ($\ln \omega(T_{ij})$ is a concave function), so that there is only one (absolute) maximum.

The final result (2.15) represents the most probable flow between each point of origin i and each point of destination j. This distribution function of the flows within the system appears to correspond to a gravity specification. Thus entropy theory appears to offer another foundation for a gravity model.

The spatial distribution of trips, represented by (2.15), can be used for several purposes. First, the result can be used to derive an 'optimal' (i.e., most likely) trip distribution in a spatial system, given O_i, D_j and C. One then has to determine first the parameters A_i, B_j and β. To calibrate model (2.15) at some base year, with data on a given set of origin zones O_i and destination zones D_j, a parameter value for the distance friction function has to be found that 'best' reproduces the known base year trip pattern T_{ij} (see, e.g., Foot, 1986). Such a (numerical) estimation can be carried out iteratively (see Hyman, 1969), for example, by starting with initial values of all B_j terms and of β, by calculating the resulting values of all A_i terms by means of (2.16), by calculating, in turn, a new series of B_j terms by means of (2.15), and so forth, until the procedure converges to an equilibrium point. By substituting the resulting values of A_i and B_j into (2.15) and, next, by substituting T_{ij} into (2.1)-(2.4), one can calibrate β, given the value C. Once A_i, B_j and β have been estimated, the "optimal" (i.e., most likely) flows T_{ij} can be determined. Clearly, if C is unknown (which is frequently the case), A_i and B_j can be estimated up to a multiplicative constant, viz., the exponential cost function. By varying the parameter β or by using prior estimates of β, different states of the system are found. In this way one might also approximate the actual state of the system for a particular value of β in the distance deterrence function.

A second use of expression (2.15) is for projecting spatial flows. In this case actual observations on T_{ij} from the past are used to estimate the unknown parameters A_i, B_j and β. The parameters A_i, B_j and β are determined by means of regression analysis or by alternative numerical methods; see, for example, Batty (1970b), Batty and Mackie (1972), Chisholm and O'Sullivan (1973), Hyman (1969), and Wilson et al. (1969). One might, for instance, take the natural logarithm of both sides of (2.15). Then, by assuming that T_{ij} can be explained in terms of O_i, D_j and c_{ij}, one can estimate the corresponding parameters. Once the parameters A_i, B_j and β are known, model (2.15) can be used as a relationship projecting the most likely future development of the spatial system concerned, given knowledge of O_i and D_j. However, the special assumption implicit in the invariance of λ_i, μ_j and β should be borne in mind; it is obvious that a change in the future structure of the system will affect the dual variables and hence the calibrated parameters.

Finally, some attention should be paid to the interpretation of parameters β, A_i and B_j in expression (2.15). Parameter β is a Lagrange multiplier and represents the marginal change in the optimal value of $\ln \omega(T_{ij})$, consequent upon a change in the transport budget C. In other words, β represents an increase in the logarithm of the number of possible states of the system if the transport budget increases.

The Lagrange multipliers λ_i and μ_j can be interpreted similarly. The parameters A_i and B_j are associated with λ_i and μ_j, respectively, according to (2.13) and (2.14). For example, assume that a particular O_i undergoes a positive change. According to (2.17) all B_j parameters decline, which leads to positive changes in all A_i parameters according to (2.16), until finally a new equilibrium is attained. Therefore, a rise in O_i leads to a positive change in all A_i parameters and to a negative change in all B_j parameters. Such a positive change in all A_i parameters denotes that a stronger pull is exerted by all points of origin upon the number of possible states of the spatial system, whereas the strongest pull is exerted by the point i itself; in (2.15) both A_i and O_i have undergone a rise. The B_j parameters can be interpreted in a similar manner. In fact, B_j and A_i represent the pull and push effects which are exerted by the points of destination and origin. They compensate for the increased attractiveness of a given point such that the additivity conditions are satisfied.

The pioneering work of Wilson (1970) generated consequently a new interest in spatial interaction analysis and stimulated many others to develop more general approaches which led to similar or adjusted models. Some of these approaches focused attention on the links between the entropy principle and mathematical programming theory (see also Chapter 3), whilst other focused on the relationships with information minimization approaches (see, e.g., Batten, 1983, and Snickars and Weibull, 1977) or the incorporation of entropy models in broader regional or transportation models. Especially in the 1970s, many discussions and new applications of entropy theory in relation to spatial interaction modelling took place. Examples can found amongst others in Batty (1970b), Berry and Schwind (1969), Bussière and Snickars (1970), Cesario (1973), Champernowne et al. (1976), Charnes et al. (1972), Cordey Hayes and Wilson (1971), Evans (1976), Lowe and Moryades (1975), Nijkamp (1975, 1976), Nijkamp and Paelinck (1974), Openshaw (1976), Scott (1971), Sheppard (1975), Webber (1976), Wilson (1970), and Van Zuylen (1979). This wave of studies has continued in the past decade.

Now that entropy theory appears to provide a new basis for the use of gravity models, one may consider whether this theory fulfils the methodological requirements mentioned in Chapter 1.

First, the results of the entropy approach guarantee that logical conditions (such as additivity conditions and distance friction conditions) are satisfied. However, the assumption that each assignment is equally likely, provided the constraints are satisfied, is less plausible.

Secondly, there is a high degree of similarity between the physical background of entropy and its use in a spatial interaction system. The closeness of the spatial system is guaranteed by conditions (2.1)-(2.4), while the notion of fixed mean energy in the system is reflected by the travel budget. The relationship between micro- and meso-states has also a firm analogy with a physical system. Clearly, this correspondence relates only to the formal structure.

Thirdly, the behavioural background of entropy or its underlying social choice mechanism is not directly clear. There is a priori no reason to believe that people behave like particles in a physical system. This problem will be considered in more detail in the next sections.

Next, the mathematical specification of (2.15) has a rationale in the context of the gravity model, but it may be worthwhile to investigate whether this specification may be related to a meaningful specification of a utility function in the entropy context (see also Sections 2.3 and 2.4).

Finally, since equation (2.15) is essentially the gravity model, its empirical value is in general satisfactory. Two problems inherent in the use of (2.15) are worth mentioning. First, the assumption of fixed unit transportation cost c_{ij} in a given travel budget C implies that capacity problems of a network are left aside. For example, a high flow T_{ij} might exceed the capacity between i and j and cause congestion, leading to an increase of c_{ij} (unless capacity constraints are specified a priori). This possibility is left aside in the entropy approach, although it might be included by means of an iterative adjustment of the travel budget and/or cost coefficient in case of congestion. Secondly, when (2.15) is used as a projection model for future flows, the calibrated parameters A_i, B_j and β are only valid for a given O_i, D_j, c_{ij} and C (as we mentioned before). Any future change in these values will affect the parameters A_i, B_j and β, so that (2.15) cannot be used as a straightforward projection model. Here again an adjustment of parameters has to take place.

2.3 Alternative Specifications of the Entropy Model

In this section we will pay attention to various other specifications of spatial interaction phenomena that lead to the same type of entropy function. It should be noted that the ultimate specification of (2.15) does not necessarily rest upon a statistical mechanics background (cf. also Smith, 1972). By means of information theory (cf. Jaynes, 1957), by means of Bayes' theorem for conditional probabilities (cf. Hyman, 1969), or by means of maximum likelihood estimators of the spatial interaction model (cf. Evans, 1971), a similar result can be obtained. All these methods, however, are essentially based on the same statistical background, so that they suffer all from the rather rigorous and restrictive assumptions implicit in the entropy approach (such as the use of a simplified travel budget, and the absence of other behavioural elements). Some of these alternative approaches will be discussed later on in Section 2.4.

Furthermore, alternative entropy specifications leading to the same entropy model have been proposed by various authors.

For instance, it has been shown by Coelho (1977) and Coelho and Wilson (1977) that the standard model (2.15) can be derived from a different formulation of problem (2.8) by including the cost constraint directly in the objective function w* as follows:

$$\text{Max } w^* = \frac{-1}{\beta} \sum_i \sum_j T_{ij} (\ln T_{ij} - 1 + \beta c_{ij})$$

subject to

$$\begin{aligned} \sum_j T_{ij} &= O_i \qquad \forall i \\ \sum_i T_{ij} &= D_j \qquad \forall j \end{aligned} \tag{2.18}$$

Obviously, the only difference here is the definition of the balancing factors A_i and B_j which are now equal to:

$$\begin{aligned} A_i &= \exp(-\beta\lambda_i) / O_i \\ B_j &= \exp(-\beta\mu_j) / D_j \end{aligned} \tag{2.19}$$

It is noteworthy that the formal solution of (2.18) is still the same as expression (2.15). The non-linear programming model (2.18) has been interpreted by Coelho (1977) as the maximization of the measure of the consumer surplus arising from travel demand subject to the origin and destination constraints. Thus the objective function in (2.18) corresponds to a utility function expressed in money units. In this framework the parameter β is regarded as a factor converting utility (expressed in money units) into entropy units; consequently entropy is interpreted as "utility in other systems of units, say 'utils'" (Coelho, 1977, p.101) and β as the marginal utility of expenditure on transport. It is interesting to note that model (2.18) may also be interpreted as a **multiple objective model** with two objective functions, viz.,:

$$\text{Max } w^* = -\alpha_1 \sum_i \sum_j T_{ij} (\ln T_{ij} - 1) - \alpha_2 \sum_i \sum_j c_{ij} T_{ij}$$

subject to

$$\sum_j T_{ij} = O_i \qquad \forall i \tag{2.20}$$

$$\sum_i T_{ij} = D_j \qquad \forall j$$

where the weight coefficients α_1 and α_2 satisfy the usual additivity conditions:

$$\alpha_1 + \alpha_2 = 1 \tag{2.21}$$

The solution of (2.20) is still a SIM of the form (2.15), where

$$\beta = \alpha_2 / \alpha_1 \tag{2.22}$$

$$A_i = \exp(-\lambda_i / \alpha_1) / O_i \tag{2.23}$$

$$B_j = \exp(-\mu_j / \alpha_1) / D_j \tag{2.24}$$

Thus, SIMs can also be reinterpreted and reformulated in the context of multiple objective programming analysis (see also Hallefjord and Jörnsten, 1986, and Nijkamp and Reggiani, 1989b), with two conflicting objectives, viz., maximization of interactivity (measured by entropy) and minimization of total interaction costs.

In the context of multiple objective programming linked to the concept of prior information (see Batten, 1983), it is interesting to observe that the following entropy formulation discussed by Leonardi (1985a) may also be considered:

$$\hat{\omega} = -\frac{1}{\beta}\sum_i \sum_j T_{ij} \left(\ln T_{ij} / O_i D_j + \beta c_{ij}\right) \tag{2.25}$$

It is easy to see that maximization of (2.25), subject to the relevant constraints, leads again to a SIM of the form (2.15).

Thus the entropy model can be derived from various alternative - but mutually connected - specifications. Another interesting generalization of SIMs is the **general theory of movement** developed by Alonso (1978). His approach includes systemic variables that are similar in structure to the balancing factors in the constrained version of the SIM. The parameters associated with these variables indicate the impact of the spatial system as a whole on place-to-place flows. Thus this means that the flows T_{ij} from i to j are not only determined by push variables in i and pull variables in j, but also by the attributes of alternative origins and destinations.

The generality of the Alonso model has been discussed in several publications (see among others Alonso, 1978, Anselin and Isard, 1979, Hua, 1980, Ledent, 1980, 1981, Merkies and Van der Meer, 1989, Nijkamp and Poot, 1987, and Wilson, 1980). In particular it has been observed that by prespecifying the values of the systemic variable parameters, Alonso's model is reduced to the family of SIMs (including unconstrained, singly-constrained and doubly-constrained interaction models as special cases) (see also Fotheringham and Dignan, 1984, and Nijkamp and Reggiani, 1988a).

In a concise way, the model proposed by Alonso can be represented by means of the following five equations:

$$T_{ij} = S_i Y_j W_i^{\sigma_i - 1} Z_j^{\tau_j - 1} f(c_{ij}) \tag{2.26}$$

$$W_i = \sum_j Y_j Z_j^{\tau_j - 1} f(c_{ij}) \tag{2.27}$$

$$Z_j = \sum_j S_i W_i^{\sigma_i - 1} f(c_{ij}) \tag{2.28}$$

$$\sum_j T_{ij} = S_i W_i^{\sigma_i} \tag{2.29}$$

$$\sum_i T_{ij} = Y_j Z_j^{\tau_j} \tag{2.30}$$

where S_i represents the propulsiveness of i and Y_j the attractiveness of j; W_i and Z_j are the balancing factors defined in equations (2.27) and (2.28), respectively; W_i enters also in the total outflow from i indicating the relative attractiveness exerted by the rest of the system as seen from i with a response rate σ_i (see equation (2.29)); analogously Z_j appears in the total inflow into j and indicates the relative unattractiveness of the rest of the system as seen from j with a response rate τ_j (see equation (2.30)).

A substantive interpretation of the systemic variables W_i and Z_j is still a theme of discussion and research (see, for example, Merkies and Van der Meer, 1989, who analyze additional assumptions regarding the Alonso model). However, it is interesting to recall the interpretation of W_i and Z_j as **locational** or **accessibility** measures (see Fotheringham and Dignan, 1984), which correspond to a similar interpretation of A_i and B_j. The inverse of A_i is usually interpreted as a measure of the accessibility of a zone or region i (see also Weibull, 1976).

It is easily seen that for some specific values of σ_i and τ_j we obtain the typical spatial interaction family; in particular, for $\sigma_i = \tau_j = 1$ we get the unconstrained spatial interaction model; for $\sigma_i = \tau_j = 0$ we obtain the doubly constrained SIM; for $\sigma_i = 0$ and $\tau_j = 1$ we have the production-constrained SIM; finally for $\sigma_i = 1$ and $\tau_j = 0$ we get the attraction-constrained SIM. Of course, by using alternative values of σ_i ($\sigma_i \geq 0$) and τ_j ($\tau_j \geq 0$), a broader class of SIMs may emerge.

It is interesting to notice that the Alonso model is consistent with the maximization of entropy in a general spatial interaction system (see for a proof Nijkamp and Reggiani, 1988a). This is in agreement with the property of the SIM as a particular case of the Alsonso model (see, e.g., Fotheringham and Dignan, 1984, and Wilson, 1980).

2.4 Alternative Theoretical Backgrounds of the Entropy Model

As mentioned before, the entropy model has its roots in statistical mechanics, and for this reason several attempts have been made to base the use of this model more rigorously on social science theories, as the macro-statistical principle of maximum entropy is not always quite clear as a **behavioural** device. The conflict between social physics and economic utility theory was reflected in studies of Arrowsmith (1973), Beckmann (1974), Beckmann and Golob (1971), Cochrane (1975), and Hansen (1972, 1975) among others. Beckmann and Golob even argued that entropy maximization is a metaphysical approach and that identical results can be achieved by utility maximization. Hansen (1972) proved that the entropy concept is a special case of the utility approach. In Nijkamp (1975) and Nijkamp and Paelinck (1974) it was demonstrated that the entropy approach can be interpreted in terms of a generalized cost function for transport behaviour. In the present section some of these attempts will briefly be discussed.

2.4.1 An economic utility approach

Beckmann (1974) has constructed an economic model of choice behaviour for trip making that is not only similar to the results of entropy analysis, but also richer in its interpretation and applicability. Beckmann assumed that a given household considers various potential residences after having accepted a job in location j. This household associates a utility index u_i with each residence i. This utility index includes all attractiveness elements of residence i except distance or accessibility. When these utility indices are normally distributed, the probability that a residence has a utility u or more is:

$$p(u) = \int_{u}^{\infty} \frac{1}{\sigma\sqrt{2\pi}} \exp -[(x-\mu) / 2\sigma^2] \, dx \qquad (2.31)$$

where μ and σ are the mean and the variance of the quality (acceptability) of housing, respectively. When the household takes into account the distance to work

(d_i), then the **net** utility index u_i^* of a household living in i is:

$$u_i^* = u_i - k\, d_i \tag{2.32}$$

where k is the conversion factor of distance into utility units. Next, Beckmann assumes that the household is able to specify a minimum achievement level for net utility $(\bar{u}_i)$ and that the housing supply in residence i is a_i. Thus the total expected number of trips generated by a residential area at location i at a distance d_i from the work place is:

$$a_i\, p(\bar{u}_i + kd_i) = a_i \exp\left[(\mu - \bar{u}_i) / \sigma\right] \exp\left[(-k / \sigma)\, d_i\right] \tag{2.33}$$

The total work trip distribution from i to j (i.e. T_{ij}) is equal to the number of persons employed in place j (i.e., b_j) times the expected residential attractiveness specified in (2.33):

$$\begin{aligned} T_{ij} &= a_i\, b_j\, p\,(\bar{u}_i + kd_{ij}) \\ &\simeq a_i\, b_j \exp\left[-(k / \sigma)\, d_{ij}\right] \end{aligned} \tag{2.34}$$

Clearly, (2.34) is formally analogous to the entropy model (2.15). Furthermore, this relationship is able to link the achievement level $\bar{u}_i$ and the variance of the utility index of housing to the trip pattern, while in addition to distance perceived costs also play a role. However, there are two problems. The formal derivation of (2.33) is not entirely clear and would deserve a more solid theoretical explanation. Furthermore, (2.34) only represents the traditional gravity model without the balancing factors from the entropy model (i.e., an unconstrained SIM).

2.4.2 A probabilistic utility approach

A more rigorous economic basis for the entropy model along similar lines was provided by Cochrane (1975). He assumes that trip-makers make decisions which provide the greatest net benefit (utility of any trip after subtraction of trip costs; cf. (2.32)) for them and that the trip distribution pattern reflects the overall probability of trips being chosen on this basis. Similarly to Beckmann's approach, Cochrane assumes that the probability of any particular possibility offering a utility u within a given range to a trip-maker is given by a single probability density function. Then the probability that a certain trip from place i to j will be made is equal to the probability that a trip from i to j offers a net benefit greater than that of a trip to any other place.

Cochrane then postulates a cumulative distribution function $\Phi(u)$ of u of the greatest among n independent trip samples from a common underlying distribution f(u):

$$\Phi(u) = \{\Phi(u)\}^n \tag{2.35}$$

Provided n is moderately large, the upper part of $\Phi(u)$ can be shown to be reasonably approximated by means of a simple exponential form:

$$\Phi(u) = \exp\{-n \exp[-\lambda (u-\mu)]\} \tag{2.36}$$

where λ is a parameter characterising the probability density function of u and where μ is the mean. Hence, the probability density function for the utility of the best trip can be determined via the differential of (2.36).

Next, two additional assumptions are made. First, the number of trip possibilities n_j to place j is a linear function of the number of opportunities D_j at place j:

$$n_j = \theta D_j \tag{2.37}$$

Secondly, the net benefit u^*_{ij} of a trip from i to j is equal to the gross benefit u_{ij} minus generalised trip costs kc_{ij} minus congestion costs g_j in the place of destination, i.e.:

$$u^*_{ij} = u_{ij} - kc_{ij} - g_j \tag{2.38}$$

By substituting n_j and u_{ij} into (2.36) the probability that u^*_{ij} will attain any particular value is:

$$\Phi_{ij}(u^*_{ij}) = \exp\{-\theta D_j \exp[-\lambda (u^*_{ij} - \mu + kc_{ij} + g_j)]\} \tag{2.39}$$

where Φ_{ij} (u^*_{ij}) is the cumulative distribution function of the net benefit from a preferred trip from i to j. Now the choice probability p_{ij} that a trip from i to j takes place is equal to the probability that this trip offers a net benefit greater than that to any other zone. This probability can be proven to be equal to:

$$p_{ij} = \frac{D_j \exp[-\lambda(kc_{ij} + g_j)]}{\sum_j D_j \exp[-\lambda(kc_{ij} + g_j)]} \tag{2.40}$$

Consequently, the expected number of trips T_{ij} from i to j is:

$$T_{ij} = \frac{O_i D_j \exp(-\lambda g_j - \lambda kc_{ij})}{\sum_j D_j \exp(-\lambda g_j - \lambda kc_{ij})} \tag{2.41}$$

where O_i is the number of trips originating from i. By defining:

$$\exp(-\lambda g_j) = B_j \tag{2.42}$$

the following doubly-constrained entropy model is obtained:

$$T_{ij} = \frac{B_j O_i D_j \exp(-\lambda kc_{ij})}{\sum_j B_j D_j \exp(-\lambda kc_{ij})} \tag{2.43}$$

Cochrane's analysis has demonstrated that a completely different point of departure may lead to the same results as the entropy idea. Therefore, the formal entropy model may be derived as the result of a utility approach which has a less 'suspect' behavioural foundation than traditional entropy theory. Thus although entropy theory may be criticized for its weak behavioural underpinning (see also Fisch, 1977), alternative behavioural approaches may lead to the same spatial interaction result.

The foregoing remarks suggest that the **physical** background of the entropy concept is not a prerequisite to arrive at an exponential gravity-type model (see also Smith, 1972), but that also a behavioural macro model may be used to derive analogous results. Furthermore, it can also be proven that **information theory, a Bayesian approach via conditional probabilities** and **maximum likelihood procedures** can be employed to determine the most probable trip distribution pattern in a spatial interaction model. These backgrounds will briefly be discussed in Subsections 2.4.3 - 2.4.5.

2.4.3 Statistical information theory

In **information theory** (see Jaynes, 1957, and Theil, 1967) an appropriate measure for the information content of a message about the realisation of an event with probability p is:

$$m = -\ln p \tag{2.44}$$

For a set of events i (i=1,...,I) the mathematical expectation of the information content is:

$$h = -\sum_i p_i \ln p_i \tag{2.45}$$

with

$$\sum_i p_i = 1 \tag{2.46}$$

so that h reflects the amount of uncertainty represented by a discrete probability distribution. Thus formulation (2.45) is the definition of entropy in information theory, given by Shannon and Weaver (1949). In the absence of any additional conditions on the probability distribution, h reaches its maximum when all probabilities are equal (according to the intuitive notion of uncertainty).

In the framework of a spatial interaction model where no prior information is available about the trip distribution (except the marginal conditions and the cost budget), it seems reasonable to employ a probability distribution v with a maximum uncertainty. Also according to the "Principle of Insufficient Reason" of Laplace, showing that when estimating a probability distribution for which there is no information available, the uniform distribution is the most plausible assumption (i.e., the distribution that has maximum uncertainty) (see also Coelho, 1977, and Taha, 1986), i.e.,

$$\max v = -\sum_i \sum_j p_{ij} \ln p_{ij} \tag{2.47}$$

where p_{ij} is the trip probability from i to j satisfying the additivity condition:

$$\sum_i \sum_j p_{ij} = 1 \tag{2.48}$$

Furthermore, the marginal conditions and the travel budget may also be introduced:

$$\sum_j p_{ij} = O_i / T \tag{2.49}$$

$$\sum_i p_{ij} = D_j / T \tag{2.50}$$

and

$$\sum_i \sum_j p_{ij} \, c_{ij} = C / T \tag{2.51}$$

It is clear that this specification gives rise to the same results as the entropy model discussed in Section 2.2. Thus the maximization of the number of micro-states of the system leads to the same results as the maximization of the expected information content of uncertain events. The only difference is that the information approach does not make use of Stirling's approximation formula.

2.4.4 Bayesian statistics

The **Bayesian** version of the gravity-type model is based on a comparison of any two alternative states of a spatial system S' and S'' with respective trip distribution probabilities p'_{ij} and p''_{ij}. The observed prior probabilities are denoted by p^*_{ij}. Then the evidence of S' above S'' is, given the prior state p^*_{ij}, equal to (see Feller, 1968, and Hyman, 1969):

$$e = T \{\sum_i \sum_j p^*_{ij} \ln p'_{ij} - \sum_i \sum_j p^*_{ij} \ln p''_{ij}\} \tag{2.52}$$

By comparing the state S' successively with each alternative state, it is clear that the most probable trip distribution pattern is found by maximizing:

$$e^* = \sum_i \sum_j p^*_{ij} \ln p'_{ij} \tag{2.53}$$

Assuming now the general spatial interaction model:

$$p'_{ij} = A' \, B' \, f(\beta, c_{ij}) \tag{2.54}$$

then the parameters of (2.54), viz. A'_i, B'_j and β, have to be calibrated such that (2.53) is at a maximum. This leads to the following conditions:

$$\sum_j p'_{ij} = \sum_j p^*_{ij} = O_i / T \tag{2.55}$$

$$\sum_i p'_{ij} = \sum_i p^*_{ij} = D_j / T \tag{2.56}$$

$$\sum_i \sum_j p'_{ij} \, c_{ij} = \sum_i \sum_j p^*_{ij} \, c_{ij} = C / T \tag{2.57}$$

The latter result is analogous to the entropy model if:

$$f\ (\beta, c_{ij}) = \exp\ (-\beta c_{ij}) \tag{2.58}$$

and if A'_i and B'_j are defined according to (2.16) and (2.17).

It is clear that a Bayesian approach to the spatial interaction model leads to the same gravity-type results, but its evident weakness is the prior specification of (2.54) and (2.58), so that the exponential and power structure of a gravity model is already imposed **a priori**.

2.4.5 Maximum likelihood approach

Finally, the **maximum likelihood** approach to the spatial interaction model may be mentioned (see Batty and Mackie, 1972, and Evans, 1971). Here again it can be shown that entropy maximization is equivalent to maximizing the likelihood of the spatial macro-state. The production-attraction model is assumed to be:

$$p_{ij} = \bar{A}_i \bar{B}_j \exp(-\beta c_{ij}) \tag{2.59}$$

Assuming that the p_{ij}'s are stochastic and independent variables and assuming a sample of observations of trip flows $\hat{T}_{ij}$ between i and j, the joint probability of the observations $\hat{n}_{ij}$ can be proven to be proportional to:

$$\prod_i \prod_j p_{ij}^{\hat{n}_{ij}} = \prod_i \prod_j [\bar{A}_i \bar{B}_j \exp(-\beta c_{ij})]^{\hat{n}_{ij}} \tag{2.60}$$

Therefore, a maximum likelihood estimator of the parameters is obtained by maximizing (2.60) subject to the additivity conditions. The parameter estimators can be proven to be equal to:

$$\bar{\bar{a}}_i = \pi^{-1} \sum_j \hat{n}_{ij} [\sum_j \bar{\bar{B}}_j \exp(-\beta c_{ij})]^{-1} \tag{2.61}$$

$$\bar{\bar{b}}_j = \pi^{-1} \sum_i \hat{n}_{ij} [\sum_i \bar{A}_i \exp(-\beta c_{ij})]^{-1} \tag{2.62}$$

where π is the sample size.

This system can be solved in an iterative way, once β is known. An additional condition from the maximum likelihood procedure appears to be:

$$\pi^{-1} \sum_i \sum_j \hat{n}_{ij} c_{ij} = \sum_i \sum_j p_{ij} c_{ij} \tag{2.63}$$

On the basis of (2.61) - (2.62) the unknown parameters (including β) can be calibrated.

It should be noted that the likelihood approach to the spatial interaction model leads to the same expression as the original entropy approach, but here again the multiplicative and exponential spatial interaction model has been assumed **a priori**.

2.5 Concluding Remarks

Our conclusion is that the meso-state probability approach and the information-theoretic approach can be regarded as more adequate foundations for the entropy model than the Bayesian and likelihood approach, because in the former approaches no specific assumptions have been made **a priori** about the mathematical specification of the spatial interaction model. The Bayesian and likelihood approach are based on a prior specification of a gravity-type interaction model, so that a result which is similar to the entropy result is not a surprise. The last two methods are essentially calibration methods for the spatial interaction model.

Another conclusion is that, despite different methodological backgrounds, the gravity-type interaction model appears to be a rather common result. One should be aware, however, that all these different methods are essentially based on similar statistical considerations and on similar assumptions (like the fixed travel budget and the absence of explicit behavioural elements).

A final conclusion is that the entropy result can, in principle, also be attained by means of a traditional utility approach, so that it may be worth investigating whether a more general class of utility models does exist which may be regarded as a welfare basis of the entropy philosophy and which allows us to interpret entropy directly in behavioural terms. This will be discussed in the next chapter.

CHAPTER 3

ENTROPY AND GENERALIZED COST MINIMIZATION MODELS AT THE MACRO LEVEL

3.1 Prologue

In Chapter 2 various links between entropy (and gravity) theory and spatial interaction behaviour have been discussed and interpreted. It turned out that the entropy concept was compatible with a behavioural interpretation of spatial interaction. In the present chapter an attempt will be made to start off from the entropy philosophy itself and to find a behavioural rationale for entropy maximisation as such.

We will start our analysis in this chapter by taking for granted that there is in principle no substantial difference between the maximization of any objective function and that of entropy. Besides normal concavity conditions, such maximization procedures should also satisfy the interpretation of dual variables by means of Lagrange multipliers, etc. Thus we will develop here an interpretative approach to derive a plausible theoretical background for the use of entropy models.

The essential idea of this approach is thus that entropy maximising models can be regarded as a specific type of programming (or optimisation) models. By confronting the general features of programming models with the specific features of entropy maximising models, an aggregate utility background of entropy can then be offered. A major analytical framework discussed in this chapter is the geometric programming approach which has a rich potential for non-linear programming models.

From this perspective we will also present some empirical results of an entropy model for freight flows vis-à-vis the results of a linear programming model, in order to test the sensitivity of different methodological frameworks. Furthermore another extension of entropy theory, such as the conditional entropy approach, will also be discussed. This chapter will then be concluded with various research questions to be tackled later on in this book. The need for a closer look at micro-behavioural approaches will be emphasized there.

3.2 Entropy and Linear Programming

As mentioned above, an entropy model is essentially a mathematical programming model. In the framework of spatial interaction theory, it makes therefore sense to look for a link between the well-known spatial assignment models (see for a review Nijkamp and Paelinck, 1976) and entropy models.

The general format of a linear trip assignment model is:

$$\min \varphi = \sum_i \sum_j c_{ij} T_{ij}$$

subject to

$$\begin{aligned} &\sum_j T_{ij} = O_i \qquad \forall i \\ &\sum_i T_{ij} = D_j \qquad \forall j \\ &T_{ij} \geq 0 \end{aligned} \tag{3.1}$$

Clearly, the structure of such a model bears a clear resemblance to that of an entropy model. Therefore, several researchers have endeavoured to find a link between (3.1) and the entropy model (see among others Coelho and Wilson, 1977, Evans, 1973, Senior and Wilson 1974, Williams, 1976, and Wilson and Senior, 1974).

A closer examination of entropy model and trip assignment models by Evans (1973) showed that the average minimum trip costs of a linear assignment model of the type (3.1) are equal to the average trip costs of an entropy model, provided the cost friction coefficient approaches infinity ($\beta \rightarrow \infty$). In this case, the optimal trip pattern is equal for both types of models. Should the linear programming model involve multiple solutions, then the entropy outcome can be proven to be one of the solutions. In other words, when $\beta \rightarrow \infty$ in a maximum entropy model, then the likelihood of observing a minimum travel cost configuration overwhelms the likelihood of any other trip configuration, so that the maximum entropy trip pattern converges in probability to the optimal trip pattern of an assignment model.

In view of these results Senior and Wilson (1974) and Wilson and Senior (1974) claimed that the linear assignment model may be regarded as a specific case of the entropy model, viz. when $\beta \rightarrow \infty$. However, this statement needs some qualifications, because it is uncertain whether the above mentioned limit case is only valid for entropy models. It is plausible that alternative - and even more general - objective functions, which are optimized subject to a cost constraint, would give rise to the same conclusion when $\beta \rightarrow \infty$.

Furthermore, one may question the meaning of $\beta \rightarrow \infty$, especially because the T_{ij}'s and β are inversely related to each other. In fact, if we examine the main entropy model (2.15), it is intuitively clear that as β increases, exp $(-\beta c_{ij})$ sharply decreases for increasing c_{ij}. In other words, as β increases, the T_{ij}'s and hence C decrease and vice versa. This monotonically decreasing relationship between total trip costs and β can also be shown analytically (see Annex 3A).

Since a rise in β will have a negative impact on C, the condition $\beta \rightarrow \infty$ may lead to a degenerate entropy model where spatial interaction is hardly possible anymore. Such an extreme cost-sensitive model would then lead to cost-minimizing solutions in which the spatial-distribution feature of an assignment model would almost be lost. Therefore, besides an extreme case interpretation of a linear assignment model, it is meaningful to find a behavioural interpretation of the entropy model itself.

In this context it is interesting to mention here a slightly different version of the optimization model (3.1), proposed by Erlander (1980) and implicit in Coelho and Wilson (1977). Here an entropy constraint:

$$- \sum_i \sum_j T_{ij} \ln T_{ij} \geq H^o \tag{3.2}$$

is added to the linear trip assignment model. In this case problem (3.1)-(3.2) has the doubly-constrained SIM (2.15) as its unique optimal solution. Entropy is considered here as a measure of interactivity, or accessibility, or dispersion in the system. H^o could be a measure of interactivity predicted from the base year observations (i.e., in accordance with an observed trip matrix T^o_{ij}). In conclusion, the objective function of the optimising mathematical programming model (3.1) is equivalent to the cost constraint in the mathematical program (2.8)-(2.9). Thus the entropy model can be interpreted as "a device for adding dispersion from the optimal base to obtain a 'most probable' prediction" (cf. Williams and Wilson, 1980, p. 216). In this respect we may also quote Erlander and Stewart (1990, p. 92): "the gravity model is at the same time a descriptive model for equilibrium and a normative model giving system optimum".

Finally it is interesting to notice that simple variants of the linear assignment problem (3.1) (such as the quadratic transportation problem and its variants) lead to slightly different versions of the spatial interaction model (or entropy model) (2.15). These latter models have been successfully applied to empirical migration data (see notably Tobler, 1983).

3.3 Entropy and Geometric Programming

In this section an alternative interpretation of entropy models, based on a duality analysis of geometric programming theory, will be offered.

A first contribution to a utility interpretation of entropy models via geometric programming theory was given by Nijkamp and Paelinck (1974), while subsequent contributions were offered by Charnes et al. (1976), Dinkel et al. (1977), Kadás and Klafszky (1976), Nijkamp (1975, 1976), and Scott and Jefferson (1977). The approach presented here is based on an unconstrained geometric interpretation. The essential idea of this approach is that entropy maximizing models - like all programming models - should have a dual which may shed more light on the shadow values (and hence on the economic valuation) of the primal entropy model. The background is thus that entropy maximizing models are essentially optimizing models. It can be shown that the specification of an entropy maximizing model is a particular case of a dual geometric programming model. Therefore, the **primal** version of a geometric programming model can be used to derive a **dual** version of the entropy model:

primal entropy model $\longrightarrow$ dual geometric model
dual entropy model $\longrightarrow$ primal geometric model

For the sake of simplicity the entropy maximizing model will be presented here in terms of relative trip proportions p_{ij} between i and j:

$$p_{ij} = T_{ij} / T \tag{3.3}$$

Now the standard entropy model (2.8)-(2.9) can easily be rewritten in the following equivalent format:

$$\max \nu = - \sum_i \sum_j p_{ij} \ln p_{ij}$$

subject to

$$\begin{aligned}
&\sum_i \sum_j p_{ij} = 1 \\
&- \frac{O_i}{T} \sum_i \sum_j p_{ij} + \sum_j p_{ij} = 0 \qquad \forall i \\
&- \frac{D_j}{T} \sum_i \sum_j p_{ij} + \sum_i p_{ij} = 0 \qquad \forall j \\
&- \sum_i \sum_j c_{ij}\, p_{ij} + \frac{C}{T} \sum_i \sum_j p_{ij} = 0 \\
&\qquad\qquad\qquad\qquad p_{ij} \geq 0
\end{aligned} \tag{3.4}$$

For the ease of presentation we will present and analyze here a case with two places of origin and two places of destination. Then this model can easily be written as:

$$\max \nu = -p_{11} \ln p_{11} - p_{12} \ln p_{12} - p_{21} \ln p_{21} - p_{22} \ln p_{22}$$

subject to

$$\begin{pmatrix} 1 & 1 & 1 & 1 \\ 1-O_1/T & 1-O_1/T & -O_1/T & -O_1/T \\ -O_2/T & -O_2/T & 1-O_2/T & 1-O_2/T \\ 1-D_1/T & -D_1/T & 1-D_1/T & -D_1/T \\ -D_2/T & 1-D_2/T & -D_2/T & 1-D_2/T \\ C/T-c_{11} & C/T-c_{12} & C/T-c_{21} & C/T-c_{22} \end{pmatrix} \begin{pmatrix} p_{11} \\ p_{12} \\ p_{21} \\ p_{22} \end{pmatrix} = \begin{pmatrix} 1 \\ 0 \\ 0 \\ 0 \\ 0 \\ 0 \end{pmatrix} \tag{3.5}$$

The total number of unknown variables of model (3.4) is I.J. Now it can be directly verified that the latter model is a specific member of the class of **dual geometric programming models** (see Duffin et al, 1967, and Nijkamp, 1972). Therefore, it may be worthwhile to derive its corresponding primal version. The **primal** geometric programming model associated with (3.12) can adopt several forms (see, e.g.,

Nijkamp, 1975). A rather compact form can easily be derived via the primal-dual geometric relationships as follows:

$$\min \psi = \left(\prod_i \prod_j x_i^{O_i/T} y_j^{D_j/T} z^{-C/T}\right)^{-1} \left(\sum_i \sum_j x_i y_j z^{-c_{ij}}\right) \tag{3.6}$$

This primal geometric programming model is an unconstrained model. It can be regarded as the dual specification of the conventional entropy model. The variables x_i, y_j and z are dual variables ('shadow prices') related to the origin constraints, the destination constraints and the travel budget, respectively. There is formally a one-to-one relationship between the primal and dual variables of a geometric programme:

$$p_{ij} = \psi^{-1} \left(\prod_i \prod_j x_i^{O_i/T} y_j^{D_j/T} z^{-C/T}\right)^{-1} \left(x_i y_j z^{-c_{ij}}\right) \tag{3.7}$$

so that p_{ij} can be interpreted as the relative proportion of the (i,j)th term from the primal geometric programme. The relationship between the primal and dual objective function at the optimum can be represented by means of the following one-to-one primal-dual condition:

$$\nu = \ln \psi \tag{3.8}$$

Given (3.7) and (3.8), there is a unique relationship between a primal and a dual geometric programming model. Consequently, both models can be transformed into one another.

Now the next analytical and more substantive question is: How can the primal geometric model (i.e., the dual entropy model) be interpreted? The shadow prices of the primal entropy model are x_i (associated with O_i), y_i (associated with D_j) and z (associated with C). The dual entropy model is composed of two parts X and Y, defined as:

$$X = \sum_i \sum_j x_i^{-O_i/T} y_j^{-D_j/T} z^{C/T} \tag{3.9}$$

and

$$Y = \sum_i \sum_j x_i \, y_j \, z^{-c_{ij}} \tag{3.10}$$

The exponents O_i/T, D_j/T and C/T in (3.9) represent the relative influence of point i, the relative influence of point j and the average travel budget, respectively. These exponents can be regarded as the weights in the geometric average of all x_i's, y_j's and z. The shadow cost variables x_i, y_j and z are related to the imputed marginal potential spatial interaction (measured in entropy terms) caused by the presence of the origin constraints, the destination constraints and the travel budget respectively. Therefore, $x^{-O_i/T}$, $y^{-D_j/T}$ and $z^{C/T}$ may be regarded as weighted shadow variables related to push effects, pull effects and the total travel budget. X is essentially a formal expression for a generalised dual gravity model in which all points of origin and of destination as well as the average trip budget are included, so that X represents the shadow costs of the weighted aggregate interaction of the total spatial system based on attraction and repulsion effects of all places and controlled by the total travel budget. This indicates that an increase in the travel budget C will lead to a higher aggregate interaction, whereas an increase in an individual O_i or D_j may exert a negative influence. In fact, $z^{C/T}$ reflects a travel costs opportunity factor, as is easily seen by writing it as:

$$z^{C/T} = \exp\,[\,(C/T)\ln z] \tag{3.11}$$

which indicates that ln z is essentially a cost opportunity coefficient.

The second term Y has also some interesting features. The individual terms of Y, defined as:

$$Y_{ij} = x_i \, y_j \, z^{-c_{ij}} \tag{3.12}$$

are related to the pairwise interactions between all points of origin and destination. Here again the variables x_i, y_j and z are related to the repulsion power of a point of origin, the attraction power of a point of destination, and the cost restriction of the travel budget, respectively. Y_{ij} may also be written as:

$$Y_{ij} = x_i \, y_j \exp(-c_{ij} \ln z) \tag{3.13}$$

which bears a close resemblance to the original gravity model, so that ln z can again be regarded as a shadow cost parameter for bridging the interzonal distances. Obviously, the pairwise interaction between i and j declines as the unit trip costs rise. Therefore, it is conceivable to interpret (3.13) as the imputed marginal costs of interaction caused by a trip from i to j, given the trip costs c_{ij}. Hence, the term Y is associated with the total shadow costs of pairwise spatial interaction between all points of origin and destination.

Minimization of (3.6) implies that the pairwise shadow push and pull costs (based on interzonal cost friction) multiplied by the aggregate shadow costs of interaction arising from the total travel budget are at a minimum. The meaning of this statement can be clarified by using (3.7) to find an expression for the ratio of (relative) trips p_{ij} and p_{ik}:

$$\begin{aligned} p_{ij} / p_{ik} &= (y_j/y_k) \, z^{-(c_{ij} - c_{ik})} \\ &= (y_j/y_k) \exp[-(c_{ij} - c_{ik}) \ln z] \end{aligned} \tag{3.14}$$

The latter result may be compared with the ratio of relative trips calculated by means of the conventional entropy model:

$$p_{ij} / p_{ik} = (B_j \, D_j / B_k \, D_k) \exp[-(c_{ij} - c_{ik})\beta] \tag{3.15}$$

The latter result is rather similar to (3.14), so that y_j is the dual expression for the pull effect of j, and ln z is the dual expression for the traditional cost friction coefficient β.

This analysis gives rise to two important conclusions. First, by means of geometric programming theory it is possible to derive the dual entropy relationships analytically. This appears to allow a more behavioural (or at least economic) interpretation in terms of shadow costs of spatial interactions. Clearly, due to the non-linearity this economic background is less straightforward than in usual transportation models, but a close connection with traditional spatial interaction theory can certainly be demonstrated. This confirms our idea that entropy theory has a macro-behavioural rationale.

Secondly, by re-writing entropy models as geometric programming models the traditional entropy approach can easily be extended with additional constraints (for example, capacity constraints). In the traditional entropy approach it is extremely difficult to add new constraints because this will normally preclude an analytical expression for the equilibrium trip pattern. By including these constraints in a geometric programming model, one may use the solution algorithms for geometric programming models to calculate in an efficient manner the optimal (equilibrium) flows especially in those cases, for which the original entropy solution no longer holds. But also in the normal entropy model, the geometric programming approach may be used to calculate the optimal trip pattern (see for an application also Kádas and Klafszky, 1976).

Next, it may be worthwhile to investigate whether some additional properties of (3.6) can be derived. First, one may analyse the marginal effect of a change in x_i upon ψ:

$$\begin{aligned}\frac{\partial \psi}{\partial x_i} &= - x_i^{-1} (O_i/T)\, X.Y + x_i^{-1}.X. \sum_j Y_{ij} \\ &= \{-(O_i/T)\, Y + \sum_j Y_{ij}\} x_i^{-1}\, .X\end{aligned} \tag{3.16}$$

Thus it is clear that $\frac{\partial \psi}{\partial x_i} > 0$, if the aggregate shadow value of spatial interaction from place i onward (i.e., $\sum_j Y_{ij}$) exceeds the weighted shadow costs of the total spatial interaction originating from i (i.e., $(O_i/T)Y$).

A similar result can be achieved for the shadow pull variable y_i. The marginal effect of z can be calculated as:

$$\frac{\partial \psi}{\partial z} = z^{-1}(C/T)\, X.Y - z^{-1}\, .X. \sum_i \sum_j c_{ij}\, Y_{ij} \tag{3.17}$$

It is easily seen that the marginal effect of z is positive, if the average travel budget (C/T) is higher than the aggregate costs of imputed spatial interaction ($\sum_i \sum_j c_{ij} Y_{ij}$).

3.4 Spatial Patterns of Entropy and Linear Programming Models

The relationship between entropy theory and (linear and geometric) programming theory has extensively been discussed in the previous sections. It may now be interesting to make a comparison between empirical results of entropy and linear programming models (cf. also Chisholm and O'Sullivan, 1973). The empirical application presented here is related to interregional commodity flows in the Netherlands, based on a data set for aggregate freight flows between 43 Dutch regions in 1968. The aim of this exercise is to examine whether entropy models and linear programming models may lead to different results, compared to the observed flows.

The data for the unit transportation costs c_{ij} between the 43 regions were derived from the known average tariffs per ton for freight flows of various commodity classes between these regions. These tariffs can be found in the standard tables for domestic freight transport. The observations on the freight flows between these 43 regions are included in Table 3.1. The total observed transportation costs appeared to be equal to 2,193 million Dfl.

Given the column and row totals, and given the unit transportation costs, the entropy results and the linear programming results were calculated. These results are included in Tables 3.2 and 3.3, respectively. The minimum linear programming results appeared to be equal to 1,395 million Dfl. The latter result indicates that the observed and the optimal total transportation costs have a high degree of difference which is of course also caused by commodity heterogeneity.

The entropy results differ considerably from the linear programming results. First, the entropy results appear to lead to a considerable dispersion of flows within the system, whereas the linear programming results appear to lead to a limited dispersion of spatial commodity flows. Next, the entropy results include a high number of non-zero flows, whereas the linear programming results include only a limited number of non-zero flows. Furthermore, cross-hauling may occur in entropy models (for example, T_{12} and T_{21}), whereas cross-hauling is formally excluded in linear transportation models. Finally, the entropy results appear to lead to an underestimation of real intraregional flows, whereas the linear programming results appear to lead to an overestimation of intraregional flows.

These results can be justified in the light of the implicit assumptions of entropy models and linear programming models. The entropy approach attempts to maximize the number of micro-states in a certain macro-state; or in dual terms: to maximize the net interaction within a spatial system. Given the latter dual interpretation it is obvious that entropy models will lead to a considerable dispersion

Table 3.1 Observed Freight Flows between 43 Regions in the Netherlands

	1	2	3	4	5	6	7	8	9	10	11	12	13	14	15	16	17	18	19	20	21	22	23	24	25	26	27	28	29	30	31	32	33	34	35	36	37	38	39	40	41	42	43	
1	1908	68	367	478	47	155	90	309	82	62	10	52	28	20	4	7	31	10	23	15	7	20	3	6	18	48	30	1	60	13	1	15	2	9	1	23	5	15	8	4	3	2	8	4066
2	39	359	58	146	10	11	10	18	38	4	8	7	8	12	3	-	19	-	7	6	4	8	4	1	6	12	16	-	10	5	-	2	1	-	-	18	8	1	5	2	1	1	-	869
3	480	120	1070	116	68	63	66	66	19	15	8	27	21	14	65	3	19	7	2	4	4	9	9	5	4	12	11	3	32	1	-	5	3	1	1	11	4	11	5	5	-	3	1	2397
4	480	130	142	2595	38	83	73	289	87	23	14	68	32	28	5	10	51	4	13	16	4	9	3	7	8	66	35	1	68	10	1	11	5	2	2	34	12	16	28	5	5	5	6	4524
5	34	4	48	24	784	482	249	17	7	18	9	30	32	21	1	8	17	9	15	6	2	20	16	4	8	87	13	-	60	79	-	10	-	-	1	16	6	6	3	2	3	5	2	2160
6	311	5	60	36	427	3880	434	30	9	54	82	112	30	34	9	3	46	3	17	13	8	78	17	11	35	79	32	2	127	18	-	18	4	3	3	9	11	11	5	4	6	3	3	6084
7	80	17	79	64	441	344	2800	66	73	26	67	36	33	24	9	6	17	7	6	7	1	33	7	7	36	30	23	-	34	7	-	6	-	4	-	3	3	12	19	5	-	1	4	4440
8	581	36	45	487	21	42	74	1979	95	115	6	48	9	8	2	3	24	9	2	6	2	7	1	3	7	27	6	-	26	16	-	3	-	1	1	4	3	23	4	8	2	3	1	3741
9	79	11	10	281	16	19	14	186	2160	70	4	117	36	11	10	2	33	4	5	7	2	19	2	12	4	16	24	1	40	27	-	6	3	1	-	13	5	1	6	5	2	1	1	3266
10	54	3	12	51	18	38	29	200	137	1594	94	209	19	12	5	1	29	2	21	3	5	3	1	8	-	39	11	-	38	5	-	6	-	3	-	6	4	11	7	3	1	4	-	2689
11	99	8	9	18	17	70	171	54	15	89	702	190	6	5	5	1	53	16	3	7	2	14	1	15	8	30	27	3	22	5	4	7	3	12	10	16	1	7	2	2	3	1	3	1739
12	100	8	15	74	109	67	52	52	135	162	182	4367	535	408	40	29	665	23	28	44	6	28	10	15	36	156	63	14	141	53	2	43	5	4	2	36	13	69	24	21	14	11	12	7883
13	36	11	15	46	34	61	62	31	89	25	21	373	6878	360	29	37	143	17	84	39	9	19	10	18	12	91	57	1	109	15	1	26	5	10	2	31	26	35	37	10	8	10	6	8938
14	47	16	8	39	37	51	33	36	18	30	12	230	288	6504	415	103	367	122	32	62	16	30	14	34	29	118	88	-	128	73	1	27	3	10	-	39	31	67	35	31	15	10	11	9260
15	10	3	8	22	4	17	10	7	18	12	5	68	76	419	1785	57	417	338	42	129	20	18	6	23	50	112	70	4	85	22	2	19	1	7	1	41	17	34	16	35	13	11	17	4072
16	13	3	2	12	17	14	24	9	6	6	3	30	47	122	69	2044	82	287	16	35	11	16	6	16	18	98	50	1	74	15	-	18	7	8	-	40	28	204	73	116	30	18	19	3708
17	44	8	11	41	63	38	42	27	43	21	23	537	88	266	158	37	6267	50	59	251	23	33	21	53	69	275	143	4	212	44	1	33	7	16	3	64	34	60	26	16	21	13	15	9249
18	28	3	19	20	19	18	13	15	10	11	7	51	72	107	214	285	167	3520	141	145	37	51	40	73	88	184	213	7	255	265	9	39	50	56	9	110	115	228	116	52	18	32	19	6901
19	35	8	2	26	11	24	22	10	13	19	1	58	63	46	34	19	113	85	4076	457	457	27	21	36	125	167	144	2	171	90	4	125	7	24	4	53	38	50	47	30	16	25	19	6804
20	23	3	3	15	12	13	30	13	11	6	4	50	56	94	165	58	448	92	325	3734	72	16	15	39	145	200	97	3	138	38	1	47	6	13	1	37	32	34	37	12	12	25	8	6185
21	15	7	3	7	3	13	7	8	5	8	8	58	29	44	15	8	44	28	308	92	640	31	20	59	191	107	135	8	137	50	1	145	7	10	6	19	19	35	21	9	7	6	8	2380
22	16	3	4	16	15	96	8	4	3	4	5	15	12	21	7	6	33	9	15	10	17	4484	187	26	89	226	188	3	78	8	3	29	2	6	3	21	11	14	7	8	7	2	5	5726
23	11	5	1	2	16	20	1	-	-	1	3	3	5	7	5	-	15	4	11	6	4	500	384	42	75	224	36	-	50	4	2	10	1	3	3	8	2	5	9	1	3	1	2	2485
24	11	4	2	10	7	5	2	4	2	1	1	8	11	13	6	4	37	16	35	24	45	50	82	2246	100	648	363	-	125	38	1	75	12	6	-	24	7	19	17	6	3	4	5	4081
25	13	1	3	19	6	19	13	4	20	7	8	32	13	22	9	7	79	13	87	197	101	99	113	86	2077	556	41	1	87	12	-	25	5	2	1	15	13	17	19	5	2	4	8	3860
26	95	17	24	94	104	157	67	29	49	60	28	159	136	187	125	90	298	133	223	264	146	398	701	768	996	12468	551	12	660	188	3	233	22	42	6	167	97	178	198	54	30	66	48	20370
27	43	15	10	43	26	27	29	17	21	11	11	58	71	52	52	34	66	58	105	94	50	163	66	324	76	475	10058	74	1359	127	16	449	22	52	15	131	53	130	92	52	26	22	38	14715
28	8	1	-	7	-	3	1	1	-	1	2	21	8	9	8	-	14	32	21	15	4	6	2	7	4	27	57	1133	238	56	13	21	2	6	2	8	7	11	6	2	4	4	3	1777
29	115	12	32	79	49	67	27	28	34	46	12	193	156	181	81	113	285	176	252	221	125	97	53	170	139	686	2923	237	22898	867	155	502	81	142	66	500	263	299	248	173	96	118	69	33067
30	44	9	4	31	47	14	12	7	7	13	4	72	41	45	39	22	48	182	147	119	49	28	10	104	72	220	363	299	1194	4823	18	136	39	33	20	329	174	174	140	23	31	28	22	9235
31	2	-	1	-	-	3	-	2	1	-	-	3	1	2	7	1	5	11	2	2	3	7	64	4	4	8	46	21	82	7	223	18	2	9	25	8	3	3	2	1	-	-	1	583
32	23	3	10	10	10	13	11	5	8	7	2	24	27	28	21	28	47	27	123	75	150	43	40	139	89	218	1059	49	617	174	10	2617	9	13	1	54	27	40	42	22	7	13	13	5947
33	3	-	-	4	1	10	2	1	-	-	2	13	6	23	5	4	26	24	7	16	4	16	4	3	7	40	40	4	112	30	2	26	3269	161	28	103	23	44	37	3	8	11	2	4125
34	3	-	-	3	2	1	1	1	4	-	3	7	9	8	20	4	18	13	12	6	2	7	2	13	8	39	47	6	122	26	21	19	143	3051	80	305	29	32	28	4	6	8	10	4123
35	22	1	-	-	5	-	-	6	-	-	-	5	7	10	18	2	6	15	1	-	4	7	4	3	4	18	98	3	65	15	9	16	18	45	1098	106	6	10	10	3	1	3	1	1646
36	40	1	8	25	27	18	13	10	9	16	7	36	42	50	31	35	72	85	48	67	27	39	27	52	36	182	182	35	357	189	9	87	76	226	159	7257	455	242	250	46	32	40	40	10684
37	12	1	8	10	10	19	2	6	2	6	4	22	40	29	15	23	41	90	37	39	16	22	11	23	22	95	96	7	164	79	-	30	18	31	13	677	4160	287	242	32	38	32	27	6539
38	17	3	3	6	15	17	11	22	3	4	4	41	59	34	41	294	79	277	38	23	23	19	12	37	31	124	198	9	233	65	6	44	38	42	4	359	361	6895	741	265	85	50	46	10690
39	17	-	6	15	10	9	10	20	1	7	4	23	40	52	25	44	24	72	46	45	13	15	13	31	57	92	82	10	149	51	-	33	23	33	12	203	245	569	7583	203	231	123	47	10285
40	10	2	2	13	7	15	10	10	5	6	4	35	29	60	34	164	61	92	32	38	7	10	8	24	19	68	58	3	101	20	1	23	9	15	4	91	49	281	209	3586	409	94	52	5771
41	9	1	1	5	12	8	6	6	3	4	2	11	17	31	7	20	21	26	16	9	5	8	6	10	14	39	52	2	88	23	1	9	17	9	5	69	33	142	320	608	4748	369	145	6935
42	6	2	3	9	13	18	19	4	6	5	3	33	22	32	8	23	38	58	23	18	9	8	6	10	12	60	47	3	109	33	3	18	20	12	3	96	55	103	186	197	379	7785	883	10381
43	13	-	3	15	14	10	4	5	5	4	2	15	11	22	16	6	13	28	31	12	16	6	11	12	10	59	47	-	59	14	-	19	3	14	1	59	34	79	107	89	153	1371	3504	5896
	5027	913	2112	5011	2580	6053	4557	3615	3253	2571	1383	7558	9146	9477	3624	3646	10379	6079	6538	6388	2151	6518	3034	4578	4836	18529	17923	1968	30881	7703	524	5049	3950	4148	1597	11213	6519	10504	11016	5756	6477	10339	5134	280276

Table 3.2 Entropy Results for Freight Flows between 43 Regions in the Netherlands

	1	2	3	4	5	6	7	8	9	10	11	12	13	14	15	16	17	18	19	20	21	22	23	24	25	26	27	28	29	30	31	32	33	34	35	36	37	38	39	40	41	42	43	
1	1551	175	652	551	65	152	359	361	109	40	17	46	15	6	2	1	11	1	1	2	-	6	2	-	1	2	-	-	-	-	-	-	-	-	-	-	-	-	-	-	-	-	-	4066
2	203	244	85	180	9	21	43	47	27	6	3	4	6	1	-	-	2	-	-	-	-	1	-	-	-	-	-	-	-	-	-	-	-	-	-	-	-	-	-	-	-	-	-	869
3	914	103	384	325	38	90	212	213	64	23	10	27	8	4	1	1	7	-	1	1	-	4	1	-	1	1	-	-	-	-	-	-	-	-	-	-	-	-	-	-	-	-	-	2397
4	621	175	261	2344	34	80	130	471	280	52	16	60	29	8	2	1	16	1	1	3	-	3	1	-	1	2	-	-	1	-	-	-	-	-	-	-	-	-	-	-	-	-	-	4524
5	72	8	30	33	475	1115	274	34	13	25	14	27	6	3	1	1	8	1	1	2	-	26	7	1	1	7	2	-	1	-	-	-	-	-	-	-	-	-	-	-	-	-	-	2160
6	202	24	85	94	1338	3138	771	96	38	65	40	76	17	8	3	2	25	2	3	6	1	74	19	3	3	19	5	-	3	1	-	1	-	-	-	-	-	1	-	1	-	-	-	6084
7	467	48	196	150	322	754	1968	236	82	87	44	40	20	11	4	2	24	2	2	6	1	22	6	1	3	5	1	-	1	-	-	1	-	-	-	-	-	1	-	1	-	-	-	4440
8	520	59	219	602	44	104	262	1285	307	111	37	58	76	14	4	3	59	2	3	7	1	7	2	1	2	4	1	-	1	-	-	1	-	-	-	-	-	1	1	1	-	-	-	3741
9	163	35	88	371	18	43	94	319	1813	90	18	57	105	26	7	4	48	4	3	7	1	3	1	1	3	4	1	-	1	-	-	1	-	-	-	1	-	2	1	1	-	-	-	3266
10	78	10	33	90	44	103	132	152	118	924	298	281	86	52	19	12	136	10	12	28	3	10	3	5	25	17	4	-	7	2	-	3	-	-	-	2	2	8	3	4	1	-	-	2689
11	42	6	18	36	32	76	85	65	31	383	445	221	39	32	14	8	82	8	23	24	3	13	3	4	20	17	4	-	5	1	-	3	-	-	-	2	1	5	2	3	1	-	-	1739
12	97	7	41	113	51	120	64	85	80	299	183	3716	361	431	146	83	1327	62	68	162	17	17	7	23	59	95	26	2	41	11	-	16	-	-	1	14	11	36	20	27	7	2	1	7883
13	25	9	11	46	10	22	28	94	126	78	28	308	7074	464	63	39	274	28	21	89	4	3	1	6	18	24	5	1	9	3	-	5	-	-	-	4	5	19	11	16	4	1	-	8938
14	12	1	5	14	5	13	17	20	36	54	26	416	524	5427	566	254	1051	144	64	108	12	8	5	15	41	66	27	4	40	18	1	17	1	-	1	28	26	81	51	68	21	5	3	9260
15	4	-	2	5	3	6	7	8	12	24	14	177	80	710	785	477	671	253	86	150	13	10	8	15	47	72	28	4	43	20	1	19	1	1	2	30	29	91	56	77	22	7	3	4072
16	2	-	1	2	1	3	4	4	7	15	8	96	52	305	456	1001	288	280	77	110	13	8	4	15	33	53	28	5	48	23	1	17	1	1	2	54	63	243	121	205	42	14	7	3708
17	20	3	8	25	14	33	33	73	57	123	58	1128	272	926	471	212	4275	157	137	384	33	24	14	41	119	178	50	6	69	21	1	30	1	1	2	32	24	85	52	72	19	6	3	9249
18	2	-	1	3	2	4	4	5	7	15	9	85	45	206	288	333	255	2217	452	352	43	25	14	54	134	186	96	29	299	141	6	73	7	4	11	173	175	774	196	129	33	15	7	6901
19	2	-	1	3	2	5	4	5	5	15	22	80	29	78	83	78	188	385	1426	378	255	73	39	225	501	772	247	25	450	104	5	253	5	4	9	137	60	230	74	29	13	5	3	6804
20	5	1	2	5	4	10	10	10	10	31	21	168	108	116	128	99	469	265	777	1724	117	55	30	107	518	465	126	13	181	53	3	105	2	2	6	72	54	201	71	36	9	5	2	6185
21	1	-	-	1	1	3	1	1	1	4	3	21	6	15	14	14	48	38	266	138	172	50	28	205	232	703	87	6	110	26	2	60	1	1	3	31	14	43	17	6	3	1	1	2380
22	9	1	4	4	32	75	23	7	2	7	7	11	2	8	5	4	18	11	40	34	27	3166	1333	97	64	448	134	6	89	17	1	28	-	1	2	10	4	15	5	2	1	-	-	5726
23	2	-	1	1	8	18	6	2	1	2	2	5	1	3	4	2	10	7	21	18	14	1292	544	59	28	265	79	3	51	10	1	16	-	-	1	6	2	7	3	1	-	-	-	2485
24	1	-	-	1	2	5	1	1	1	5	3	23	7	15	12	13	47	38	186	100	163	146	92	701	186	1265	378	16	329	51	3	162	2	2	8	33	15	39	18	6	3	1	-	4081
25	2	-	1	2	2	4	4	4	4	23	14	51	18	36	33	24	119	83	363	424	161	84	38	163	788	959	88	7	111	32	2	55	1	1	3	41	19	57	26	10	4	1	1	3860
26	4	1	2	4	16	37	10	6	7	22	17	115	34	82	72	55	263	164	797	543	696	845	515	1581	1366	10281	1116	43	647	160	10	352	5	7	20	127	63	169	78	26	11	4	2	20370
27	1	-	-	1	3	6	2	1	1	3	2	22	5	24	20	21	50	59	179	103	61	177	108	331	88	783	7307	139	3702	483	28	488	14	21	46	194	59	79	36	12	4	2	1	14715
28	-	-	-	-	-	-	-	-	-	-	-	3	1	4	4	5	7	23	28	18	5	10	5	18	9	39	178	292	638	214	22	56	8	14	33	78	23	27	12	4	2	1	-	1777
29	1	-	-	1	2	6	2	2	2	9	6	52	13	52	46	53	104	277	490	223	115	176	104	432	166	682	5558	750	17414	2350	125	1585	52	69	173	1083	218	326	152	43	21	7	3	33067
30	-	-	-	-	1	1	1	1	1	3	2	17	6	28	25	29	37	153	133	77	32	40	25	78	57	198	850	295	2756	2300	49	364	44	27	68	926	252	174	106	23	17	7	3	3235
31	-	-	-	-	-	-	-	-	-	-	-	1	-	1	1	1	2	8	6	4	2	2	1	5	3	12	46	27	134	45	83	14	11	18	58	57	16	12	9	2	1	1	-	583
32	1	-	-	1	1	2	1	1	1	5	3	22	8	25	22	20	49	74	302	142	68	61	35	233	90	407	804	72	1738	340	14	978	9	11	28	124	68	103	45	14	6	2	1	5947
33	-	-	-	-	-	-	-	-	-	-	-	1	-	1	1	2	1	7	6	3	1	1	1	3	2	6	23	10	56	41	11	9	2742	894	93	98	24	21	18	3	2	1	-	4125
34	-	-	-	-	-	-	-	-	-	-	-	-	-	1	1	1	1	4	4	2	1	1	1	3	2	7	32	16	68	23	17	10	815	2821	166	48	15	10	8	2	1	-	-	4123
35	-	-	-	-	-	-	-	-	-	-	-	1	-	2	2	3	3	11	10	7	3	4	3	8	4	21	70	39	174	58	55	25	87	169	699	100	30	25	19	4	2	1	-	1646
36	1	-	-	-	-	1	1	1	1	3	2	15	5	32	27	52	41	139	130	77	20	18	10	38	54	116	255	80	944	688	46	98	79	43	87	5045	1200	534	577	89	62	28	13	10684
37	-	-	-	-	-	-	-	1	1	2	1	13	7	31	28	64	33	150	60	61	13	8	5	18	27	62	81	25	201	199	13	58	21	14	28	1272	1871	1129	786	118	74	37	16	6539
38	1	-	-	1	1	2	2	2	3	8	5	43	27	101	91	253	121	679	237	234	42	29	14	49	81	168	113	30	309	140	10	89	18	10	24	580	1156	4315	1208	290	114	57	24	10690
39	-	-	-	-	1	1	1	1	1	3	2	22	14	57	50	112	66	153	68	74	15	9	6	20	33	70	45	12	128	76	7	34	14	7	16	559	718	1078	5491	607	453	161	74	10285
40	1	-	-	1	-	1	1	1	2	6	3	33	23	86	78	217	104	115	31	42	7	3	2	8	15	27	17	4	41	19	2	13	3	2	4	58	122	294	690	3119	364	102	55	5771
41	-	-	-	-	-	-	-	-	1	2	1	12	7	34	29	58	35	38	17	13	4	2	1	4	7	14	8	3	27	18	2	7	2	1	3	89	100	150	670	473	3877	802	355	6935
42	-	-	-	-	-	-	-	-	-	-	-	4	2	9	9	20	13	18	7	7	1	1	-	2	2	6	4	1	10	8	1	3	1	-	1	42	52	78	246	137	831	7038	1640	10381
43	-	-	-	-	-	-	-	-	-	-	-	2	1	8	6	13	6	10	4	4	1	-	-	1	1	3	2	1	6	4	-	2	-	-	1	24	27	41	139	92	454	2023	2915	5896
	5027	913	2112	5011	2580	6053	4557	3615	3253	2571	1383	7558	9146	9477	3624	3646	10379	6079	6538	6388	2151	6518	3034	4578	4836	18529	17923	1968	30881	7703	524	5049	3950	4148	1597	11713	6519	10504	11016	5756	6477	10339	5134	280276

Table 3.3 Linear Programming Results for Freight Flows between 43 Regions in the Netherlands

	1	2	3	4	5	6	7	8	9	10	11	12	13	14	15	16	17	18	19	20	21	22	23	24	25	26	27	28	29	30	31	32	33	34	35	36	37	38	39	40	41	42	43	
1	4065	-	-	-	-	-	-	-	-	-	-	-	-	-	-	-	-	-	-	-	-	-	-	-	-	-	-	-	-	-	-	-	-	-	-	-	-	-	-	-	-	-	-	4066
2	-	869	-	-	-	-	-	-	-	-	-	-	-	-	-	-	-	-	-	-	-	-	-	-	-	-	-	-	-	-	-	-	-	-	-	-	-	-	-	-	-	-	-	869
3	444	-	1952	-	-	-	-	-	-	-	-	-	-	-	-	-	-	-	-	-	-	-	-	-	-	-	-	-	-	-	-	-	-	-	-	-	-	-	-	-	-	-	-	2397
4	-	-	-	4523	-	-	-	-	-	-	-	-	-	-	-	-	-	-	-	-	-	-	-	-	-	-	-	-	-	-	-	-	-	-	-	-	-	-	-	-	-	-	-	4524
5	-	-	-	-	2160	-	-	-	-	-	-	-	-	-	-	-	-	-	-	-	-	-	-	-	-	-	-	-	-	-	-	-	-	-	-	-	-	-	-	-	-	-	-	2160
6	-	-	-	-	30	6053	-	-	-	-	-	-	-	-	-	-	-	-	-	-	-	-	-	-	-	-	-	-	-	-	-	-	-	-	-	-	-	-	-	-	-	-	-	6084
7	-	-	-	-	-	-	4440	-	-	-	-	-	-	-	-	-	-	-	-	-	-	-	-	-	-	-	-	-	-	-	-	-	-	-	-	-	-	-	-	-	-	-	-	4440
8	-	-	43	83	-	-	-	3615	-	-	-	-	-	-	-	-	-	-	-	-	-	-	-	-	-	-	-	-	-	-	-	-	-	-	-	-	-	-	-	-	-	-	-	3741
9	-	-	-	13	-	-	-	-	3253	-	-	-	-	-	-	-	-	-	-	-	-	-	-	-	-	-	-	-	-	-	-	-	-	-	-	-	-	-	-	-	-	-	-	3266
10	-	-	117	-	-	-	-	-	-	2571	-	-	-	-	-	-	-	-	-	-	-	-	-	-	-	-	-	-	-	-	-	-	-	-	-	-	-	-	-	-	-	-	-	2689
11	194	44	-	-	-	-	-	-	-	-	1383	-	-	-	-	-	-	-	-	-	-	-	-	-	-	-	-	-	-	-	-	-	-	-	-	-	-	-	-	-	-	-	-	1739
12	324	-	-	-	-	-	117	-	-	-	-	7558	-	-	-	-	-	-	-	-	-	-	-	-	-	-	-	-	-	-	-	-	-	-	-	-	-	-	-	-	-	-	-	7883
13	-	-	-	-	-	-	-	-	-	-	-	-	8938	-	-	-	-	-	-	-	-	-	-	-	-	-	-	-	-	-	-	-	-	-	-	-	-	-	-	-	-	-	-	8938
14	-	-	-	-	-	-	-	-	-	-	-	-	-	9258	-	-	-	-	-	-	-	-	-	-	-	-	-	-	-	-	-	-	-	-	-	-	-	-	-	-	-	-	-	9260
15	-	-	-	-	-	-	-	-	-	-	-	-	-	218	3562	-	291	-	-	-	-	-	-	-	-	-	-	-	-	-	-	-	-	-	-	-	-	-	-	-	-	-	-	4072
16	-	-	-	-	-	-	-	-	-	-	-	-	-	-	62	3646	-	-	-	-	-	-	-	-	-	-	-	-	-	-	-	-	-	-	-	-	-	-	-	-	-	-	-	3708
17	-	-	-	-	-	-	-	-	-	-	-	-	-	-	-	-	9248	-	-	-	-	-	-	-	-	-	-	-	-	-	-	-	-	-	-	-	-	-	-	-	-	-	-	9249
18	-	-	-	392	-	-	-	-	-	-	-	-	-	-	-	-	430	6079	-	-	-	-	-	-	-	-	-	-	-	-	-	-	-	-	-	-	-	-	-	-	-	-	-	6901
19	-	-	-	-	-	-	-	-	-	-	-	-	-	-	-	-	-	-	6137	93	-	-	-	-	574	-	-	-	-	-	-	-	-	-	-	-	-	-	-	-	-	-	-	6804
20	-	-	-	-	-	-	-	-	-	-	-	-	-	-	-	-	-	-	-	6184	-	-	-	-	-	-	-	-	-	-	-	-	-	-	-	-	-	-	-	-	-	-	-	6185
21	-	-	-	-	-	-	-	-	-	-	-	-	-	-	-	-	-	-	-	-	2151	-	-	-	229	-	-	-	-	-	-	-	-	-	-	-	-	-	-	-	-	-	-	2380
22	-	-	-	-	-	-	-	-	-	-	-	-	-	-	-	-	-	-	-	-	-	5726	-	-	-	-	-	-	-	-	-	-	-	-	-	-	-	-	-	-	-	-	-	5726
23	-	-	-	-	-	-	-	-	-	-	-	-	-	-	-	-	-	-	-	-	-	792	1693	-	-	-	-	-	-	-	-	-	-	-	-	-	-	-	-	-	-	-	-	2485
24	-	-	-	-	-	-	-	-	-	-	-	-	-	-	-	-	-	-	-	-	-	-	-	4081	-	-	-	-	-	-	-	-	-	-	-	-	-	-	-	-	-	-	-	4081
25	-	-	-	-	-	-	-	-	-	-	-	-	-	-	-	-	-	-	-	-	-	-	-	-	3859	-	-	-	-	-	-	-	-	-	-	-	-	-	-	-	-	-	-	3860
26	-	-	-	-	390	-	-	-	-	-	-	-	-	-	-	-	-	-	-	-	-	-	1277	-	174	18529	-	-	-	-	-	-	-	-	-	-	-	-	-	-	-	-	-	20370
27	-	-	-	-	-	-	-	-	-	-	-	-	-	-	-	-	-	-	-	-	-	-	-	-	-	-	14715	-	-	-	-	-	-	-	-	-	-	-	-	-	-	-	-	14715
28	-	-	-	-	-	-	-	-	-	-	-	-	-	-	-	-	-	-	-	-	-	-	-	-	-	-	-	1777	-	-	-	-	-	-	-	-	-	-	-	-	-	-	-	1777
29	-	-	-	-	-	-	-	-	-	-	-	-	-	-	-	-	-	-	-	-	-	-	-	-	-	-	3208	-	29858	-	-	-	-	-	-	-	-	-	-	-	-	-	-	33067
30	-	-	-	-	-	-	-	-	-	-	-	-	-	-	-	-	-	-	-	-	-	-	64	-	-	-	-	83	1023	7703	-	-	-	-	-	361	-	-	-	-	-	-	-	923
31	-	-	-	-	-	-	-	-	-	-	-	-	-	-	-	-	-	-	-	-	-	-	-	-	-	-	-	59	-	-	524	-	-	-	-	-	-	-	-	-	-	-	-	58
32	-	-	-	-	-	-	-	-	-	-	-	-	-	-	-	-	-	-	401	-	-	-	-	497	-	-	-	-	-	-	-	5049	-	-	-	-	-	-	-	-	-	-	-	594
33	-	-	-	-	-	-	-	-	-	-	-	-	-	-	-	-	-	-	-	-	-	-	-	-	-	-	-	-	-	-	-	-	3950	26	-	-	-	-	-	-	-	-	-	412
34	-	-	-	-	-	-	-	-	-	-	-	-	-	-	-	-	-	-	-	-	-	-	-	-	-	-	-	-	-	-	-	-	-	4122	-	149	-	-	-	-	-	-	-	412
35	-	-	-	-	-	-	-	-	-	-	-	-	-	-	-	-	-	-	-	-	-	-	-	-	-	-	-	49	-	-	-	-	-	-	1597	-	-	-	-	-	-	-	-	164
36	-	-	-	-	-	-	-	-	-	-	-	-	-	-	-	-	-	-	-	-	-	-	-	-	-	-	-	-	-	-	-	-	-	-	-	10683	-	-	-	-	-	-	-	1068
37	-	-	-	-	-	-	-	-	-	-	-	-	-	-	-	-	-	-	-	-	-	-	-	-	-	-	-	-	-	-	-	-	-	-	-	20	6519	-	-	-	-	-	-	653
38	-	-	-	-	-	-	-	-	-	-	-	-	-	-	-	-	-	-	-	111	-	-	-	-	-	-	-	-	-	-	-	-	-	-	-	-	-	10504	-	-	-	-	-	1069
39	-	-	-	-	-	-	-	-	-	-	-	-	-	-	-	-	-	-	-	-	-	-	-	-	-	-	-	-	-	-	-	-	-	-	-	-	-	-	10284	-	-	-	-	1028
40	-	-	-	-	-	-	-	-	-	-	-	-	-	-	-	-	-	-	-	-	-	-	-	-	-	-	-	-	-	-	-	-	-	-	-	-	-	-	-	5756	-	-	-	577
41	-	-	-	-	-	-	-	-	-	-	-	-	119	-	-	-	339	-	-	-	-	-	-	-	-	-	-	-	-	-	-	-	-	-	-	-	-	-	-	-	6477	-	-	693
42	-	-	-	-	-	-	-	-	-	-	-	-	-	-	-	-	41	-	-	-	-	-	-	-	-	-	-	-	-	-	-	-	-	-	-	-	-	-	-	-	-	10339	-	1038
43	-	-	-	-	-	-	-	-	-	-	-	-	-	-	-	-	30	-	-	-	-	-	-	-	-	-	-	-	-	-	-	-	-	-	-	-	-	-	732	-	-	-	5134	589
	5027	913	2112	5011	2580	6053	4557	3615	3253	2571	1383	7550	9146	9477	3624	3646	10379	6079	6538	6388	2151	6518	3034	4578	4836	18529	17923	1968	30881	7703	524	5049	3950	4148	1597	11213	6519	10504	11016	5756	6477	10339	5134	2802

of spatial flows. Even the intraregional flows appear to be underestimated in the entropy approach. It is clear that such a high dispersion of flows is possible due to the fact that the travel budget is not at its minimum.

The linear programming approach however, is based on a minimization of aggregate transportation costs. Since the intraregional unit transportation costs are rather low, it is obvious that a linear transportation model will lead to rather high intraregional transportation flows, even to such a degree that the real intraregional flows are overestimated. Furthermore, cross-hauling is not allowed in linear programming models; in reality, cross-hauling is only unlikely if the commodity category in question is homogeneous. For heterogenous commodity classes like in our example, cross-hauling is in reality more probable, so that in this case entropy models might give more adequate results for interregional flows.

The correlation coefficient between observed flows and entropy flows appears to be equal to 0.92, whereas the correlation coefficient with respect to linear programming results is 0.97. These relatively high values suggest that the average fit with respect to a regression line between observed and calculated flows is rather good, albeit that the individual values may have considerable deviations.

The entire spatial picture leads to the conclusion that entropy tends to lead to an underestimation, and linear programming to an overestimation of intraregional flows. For interregional flows, however, the entropy outcomes appear to overestimate the observed flows, whereas the linear programming outcomes appear to underestimate the observed flows. Both methods appear to lead to considerable prediction errors, which can be mainly explained from the heterogeneity of the commodities in question.

3.5 Entropy Revisited

The entropy models discussed in the foregoing sections can be extended in several ways. One possible approach is to introduce the notion of a **continuous** entropy which is related to continuous situations in which no discrete spatial points are distinguished (e.g., a distribution of population through a continuous space). Continuous entropy models were studied among others by Batty (1974), Cesario and Zerdy (1975), Hobson and Cheng (1973), Ingels (1971), and Jaynes (1973).

Another extension of entropy theory is to introduce the notion of a **conditional** entropy. Assume a trip probability p_{ij}. Then a conditional probability is defined as the probability that a trip will terminate in place j, given the condition that this trip originates from place i (cf. Pluym, 1979, and Theil, 1967). The entropy function associated with such a conditional probability is:

$$w = - \sum_i \sum_j (p_{ij} / p_{i.}) \ln (p_{ij} / p_{i.}), \tag{3.18}$$

where the marginal probability p_i is defined as:

$$p_{i.} = \sum_j p_{ij} \tag{3.19}$$

Next, the expected (average) conditional entropy of the trips attracted by all destination zones, given the known prior probabilities for the total trips from all points of origin, is:

$$\begin{aligned} w^* &= - \sum_i p_{i.} \sum_j (p_{ij} / p_{i.}) \ln (p_{ij} / p_{i.}) \\ &= - \sum_i \sum_j p_{ij} \ln (p_{ij} / p_{i.}) \end{aligned} \tag{3.20}$$

It is easily seen that the latter format corresponds again to the objective function of a dual geometric model, so that problems related to (3.20) can be solved by means of standard algorithms for geometric programming problems.

This approach can be generalised to any other prior distribution (see among others, Batty and March, 1976, Bussière and Snickars, 1970, Cesario and Zerdy, 1975, Hobson, 1971, Hobson and Cheng, 1973, Good, 1963, Jaynes, 1968, Kadás and Klafszky, 1976, Kullback, 1959, Lindley, 1972, Nijkamp, 1976, and Snickars and Weibull, 1977). This implies that $p_{i.}$ may be replaced by any prior probability on certain flows which may be known, so that this prior probability is a first guess of trip probabilities (a hypothetical estimation) on the basis of whatever information is available, prior to the introduction of new information from the spatial interaction model. In this respect, Hobson has shown that:

$$w^o = \sum_i \sum_j p_{ij} \ln (p_{ij} / \hat{p}_{ij}) \tag{3.21}$$

is a unique measure of the information content (in entropy terms) in a posterior probability assignment p_{ij}, given a prior probability $\hat{p}_{ij}$.

An application of the latter approach is contained in Nijkamp (1976), in which commuting flows were assessed by means of a prior probability entropy model. The prior probabilities of commuting flows were derived from a partial behavioural trip model, in which trips were explained on the basis of the environmental and residential characteristics in the places of origin, and the employment characteristics at the destinations. By introducing these prior probabilities in the posterior entropy model, the entropy model itself could be used as a consistent spatial interaction framework to assess the posterior trip probabilities.

3.6 Concluding Remarks

In the present chapter we have essentially shown, by means of linear and geometric programming, the integration of behavioural models and entropy models which can be regarded as an extremely important means to solve the behavioural-physical dilemma in 'social physics'.

Empirical research however, has not awaited solid methodological justifications. In the past years a wide variety of entropy models have been used in many empirical contexts: budget allocation models, consumer expenditure models, trip assignment models, input-output models, migration and commuting models, transportation models, location models and so forth.

Further discussions on the empirical validity of entropy models can be found among others in Batty and Mackie (1972), Cripps and Cater (1971), Van Est and Van Setten (1979), Hathaway (1975), Kirby (1974), Openshaw (1976), Senior and Wilson (1974), and Wilson and Bennett (1985). In these studies the calibration problems of entropy models are also set out in more detail.

In retrospect, we come to the conclusion that the idea of entropy has opened a rich field of scientific research, despite the fact that the meaning of entropy varies substantially in the field of social sciences. For example, entropy has sometimes been used as a background idea for studying the stability and equilibrium of a system (see for instance, the path finding study of Georgescu-Roegen (1971) on the ecological basis and constraints of economic systems). Next, entropy has also been employed as an operational tool for solving allocation and assignment problems.

With respect to the latter issue, entropy constitutes the foundation for the traditional gravity model. For both gravity and entropy models a set of methodological criteria may be specified in order to judge the plausibility of these methods in social sciences. Both methods appear to have a behavioural economic-oriented interpretation, leading to the conclusion that from a methodological point of view the use of gravity and entropy models is logically sound and plausible. In addition, the use of these models is supported by empirical results.

On the other hand, gravity and entropy theory as a tool in social physics does not provide a neutral device for solving allocation and assignment problems. Both theories are related to a rather specific utility background as is shown by the use of geometric programming theory. Therefore, it is reasonable to conclude that gravity and entropy theory are only one particular class of spatial interaction theories. Entropy models may be extended with behavioural relationships in order to overcome the restrictive assumptions of these models (the fixed travel budget, e.g.). This leads directly to the use of geometric programming methods. A complementary approach may be to include the results of a behavioural approach in the form of a prior trip probability in a Bayesian type of entropy model. Therefore, there is a need for a better integration of behavioural models and social physics models. In this respect, we may quote Beckmann (1974): "A good utilitist will not be outdone by an entropist".

In the light of these observations it becomes also necessary to investigate the micro-utility basis of SIMs, including entropy models. Despite their macro-economic plausibility and justification, the need for an individual behavioural foundation of such models based on individual utility theory is clear. This issue will be discussed in the next chapter.

Annex 3A

Relationships between Total Trip Costs and the Cost Friction Coefficient

In this annex we will show that the trip flows T_{ij}'s (and hence the total trip costs C) and the cost friction coefficient β are inversely related to each other.

This can easily be seen by differentiating the standard entropy result (2.15) with respect to β:

$$\partial T_{ij} / \partial \beta = (A_i^* + B_j^* - c_{ij})T_{ij} \tag{3A.1}$$

with:

$$A_i^* = A_i^{-1} \partial A_i / \partial \beta \tag{3A.2}$$

and:

$$B_j^* = B_j^{-1} \partial B_j / \partial \beta \tag{3A.3}$$

It can also be derived by means of the additivity conditions on the total trips (1.5) and (1.6) that:

$$\sum_i \partial T_{ij} / \partial \beta = 0 \tag{3A.4}$$

and:

$$\sum_j \partial T_{ij} / \partial \beta = 0, \tag{3A.5}$$

so that:

$$\sum_i \sum_j B_j^* (A_i^* + B_j^* - c_{ij}) T_{ij} = 0 \tag{3A.6}$$

and

$$\sum_i \sum_j A_i^* (A_i^* + B_j^* - c_{ij}) T_{ij} = 0 \tag{3A.7}$$

Now it can easily be seen that the total trip costs of a spatial interaction system are affected by a change in β as follows:

$$\sum_i \sum_j c_{ij} \, \partial T_{ij} / \partial \beta = \sum_i \sum_j c_{ij} \, T_{ij} \, (A_i^* + B_j^* - c_{ij}) \tag{3A.8}$$

By subtracting (3A.6) and (3A.7) from (3A.8), one finds:

$$\sum_i \sum_j c_{ij} \, \partial T_{ij} / \partial \beta = - \sum_i \sum_j T_{ij} \, (A_i^* + B_j^* - c_{ij})^2 \leq 0 \tag{3A.9}$$

so that there is a monotonically decreasing relationship between total trip costs and β.

CHAPTER 4

SPATIAL INTERACTION MODELS AND UTILITY MAXIMIZING BEHAVIOUR AT THE MICRO LEVEL

4.1 Prologue

In Chapter 3 it has been demonstrated that the family of SIMs can be derived from different formulations of an entropy (or utility) maximizing macro approach and hence viewed as an optimum system's solution. In the present chapter the attention will be devoted to a further investigation into the connections between SIMs and micro-economic theory.

In this context we will show that SIMs can emerge from the maximization of a micro (or individual) choice function in both deterministic and random utility theory. In particular, the latter analysis will illustrate that a SIM is formally compatible with a multinomial logit model, pertaining to the class of well-known discrete choice models. Thus this analogy offers again a behavioural interpretation to aggregate SIMs on the basis of individual utility theory.

In order to show also the potential of SIMs from an empirical point of view, with particular attention for the utility function concerned, also an empirical application will be presented in relation to a modal choice problem for a transport system, i.e., the Milano-Bergamo corridor in Italy.

It appears that the specification of the utility function - together with the type of data used (micro and macro) - seems critical for a better understanding of the explanatory and descriptive power of SIMs.

4.2 Spatial Interaction Behaviour and Individual Choice Behaviour: Theory

4.2.1 Introduction

Both spatial interaction models and individual choice models are flow models aiming at portraying and forecasting spatial choice patterns and processes in complex geographical systems (see Fischer et al., 1990, Van Lierop and Rima, 1985, and Reggiani, 1990). In general, such flow models can be typified according to different angles (see Van Lierop and Nijkamp, 1991):

(1) the level of **aggregation**; in recent years, there has been a gradual shift from macro/meso type of approaches to micro-based analyses (see, for instance, Harsman and Snickars, 1975, McFadden, 1978, and Rouwendal, 1989).

(2) the **scope** of analysis; according to Manheim (1979), choice processes are multidimensional and can be investigated by preference and perception methods, based on stated preference or revealed preference techniques (cf. Nijkamp and Reichman, 1988).

(3) the influence of **time**; choice processes have a time component (e.g., because of learning effects) (see also Clark and Smith, 1982, De Palma and Ben-Akiva, 1981, and Weibull, 1978); this has also led to the popularity of panel and longitudinal analysis (see also Golob et al., 1984), as well as to the use of dynamic approaches.

Thus the 'geography of movement' (Lowe and Moryades, 1975) can be studied in many ways, as is also reflected in the great variety of different spatial choice and interaction models. Examples are:

A) **aggregate models**

- programming models
- gravity and entropy models
- catastrophe, bifurcation and chaos models

B) **disaggregate models**

- micro-simulation models
- conventional utility-maximization models
- random utility (or discrete choice) models
- psychometric behavioural models
- activity-based choice models
- search models

Each of these models has its own merits and its own disadvantages. The choice for a given class of models will often depend on the objectives of the analysis and on the data base (cf. Nijkamp et al., 1985). Nevertheless, it is interesting to observe that there has been a gradual shift from macro to micro approaches in the area of spatial choice and interaction analysis. Therefore, the linkages between conventional SIMs and more recently developed disaggregate models have to be investigated in more detail.

Compared to macro-oriented approaches, micro models have the following general advantages (see also Clark, 1983, Harsman and Snickars, 1975, Van Lierop and Nijkamp, 1980, 1982, McFadden, 1978, and Van Wissen and Rima, 1988):

(a) a closer orientation toward **behavioural** approaches (see for instance, Burnett, 1973, Clark and Cadwallader, 1973, Downs, 1970, Golledge and Brown, 1967, Gould, 1973, Rushton, 1969, and Saarinen, 1976);

(b) a more precise description of actual spatial interactions which may take place at **various aggregation levels** (see, for instance, Stopher et al., 1981);

(c) a better possibility for analysing choice processes on a **longitudinal, event-history or dynamic basis** (see for instance, Coleman, 1981, Halperin, 1985, Koppelman and Pas, 1985, and Tuma and Hannan, 1984);

(d) a greater **flexibility in specifying choice processes** compared to traditional approaches, without making stringent assumptions regarding equilibrium, competition, or homogeneous land use (see also De Palma and Ben-Akiva, 1981, McDonald, 1979, and Smith and Clark, 1982);

(e) a more effective way of testing the **statistical validity of empirical results from surveys or questionnaires** (see for instance, Hensher and Johnson, 1981, and Manski and McFadden, 1981);

(f) a better way of including **qualitative** information on spatial choice processes in explanatory models (see Wrigley, 1984);

(g) a more satisfactory representation of **public policy impacts** on micro spatial choice processes (see Van Lierop and Rima, 1985, Quigley, 1979, and Vollering, 1991);

(h) a more adequate representation of the **dependence** of the utility of an actor on the decision of all other actors including agglomeration and congestion effects (see Miyao and Shapiro, 1981).

Clearly, there is also a price to be paid for using disaggregate choice models: the complexity of model design and calibration increases, while also the computational costs may become fairly high.

In this context we will pay attention, in the present chapter, to the linkages between SIMs and two particular classes of disaggregate models, viz. the conventional utility maximization models (based on deterministic utility theory) and the discrete choice models (based on random utility theory).

4.2.2 Spatial interaction models and deterministic utility theory

In the framework of decision theory the conventional assumption is that individuals choose rationally among the options available to them, subject to a certain number of constraints. However, the individuals behave differently and even the same individual may behave in a different way in each point in time. Thus it is necessary to distinguish between a homogeneous population (where all members perceive utility in a similar way) and a heterogeneous population (where dispersion with respect to the mean value of the utility will be large). Obviously if the population is really homogeneous, the totality of choices will be concentrated on alternative k which corresponds to the maximum utility V_k, by neglecting all the other alternatives $j \neq k$ for which $V_j < V_k$. In other words, we will then have the following solution:

$$V_k = \max_j V_j \qquad j,k = 1,\ldots,J \tag{4.1}$$

which corresponds to deterministic utility maximization (characterizing inter alia most neo-classical urban economic models; see also Bertuglia et al., 1987, and Leonardi, 1985a).

In this framework it is interesting to indicate that the transportation problem (3.1), discussed in Chapter 3, can be reinterpreted in the context of deterministic utility theory by replacing the objective function in (3.1) as follows:

$$\max U = \sum_i \sum_j V_{ij} T_{ij} \tag{4.2}$$

where now the objective function U is based on aggregate utilities V_{ij} as well as on aggregate trips T_{ij}. In particular V_{ij} is defined as follows:

$$T_{ij} = \sum_n V_{ij}^n / T \qquad n = 1,...,T \tag{4.3}$$

where V_{ij}^n represents the utility for trip-maker n in going from zone i to zone j.

T_{ij} is then defined as follows:

$$V_{ij} = \sum_n t_{ij}^n \qquad n = 1,...,T \tag{4.4}$$

and t_{ij}^n is a dummy variable with the following characteristics:

$$t_{ij}^n = \begin{cases} 1 \text{ if trip-maker n decides to go from i to j} \\ 0 \text{ otherwise} \end{cases}$$

Consequently, we also have the constraint:

$$\sum_i \sum_j t_{ij}^n = 1 \qquad n = 1,...,T \tag{4.5}$$

Thus it can be shown (see Erlander and Stewart, 1990) that the following optimization problem holds for trip-maker n:

$$\max U = \sum_i \sum_j V_{ij}^n t_{ij}^n$$
$$\sum_i \sum_j t_{ij}^n = 1 \qquad t_{ij}^n \in (0,1) \tag{4.6}$$

This problem leads to an equilibrium solution in a perfectly competitive market and can be traced back to the minimum cost problem (3.1) which gives rise (by including constraint (3.2)) to an SIM, as we noticed in Section 3.2. Consequently, a behavioural context underlying the class of SIMs emerges again, if we simply aggregate and approximate such a deterministic utility model for individual behaviour.

4.2.3 Spatial interaction models and random utility theory

4.2.3.1 Basic concepts of random utility theory

The previous analysis dealt with the relationships between a deterministic utility model and a spatial (or entropy) model. However, it seems more realistic to analyze the behaviour of a heterogeneous population where variations around the mean utility value are large. These kinds of variations are well classified in Domencich and McFadden (1975) and broadly presented in De La Barra (1990); they can be summarized as follows:

(a) intra-individual variations

(b) intra-option variations

(c) different information about the attributes of each option among the individuals within a group

(d) mis-specification of the utility model in the model itself

(e) the variation of the exact spatial location of individuals within an aggregate group/zone (both with respect to the origin of an interaction as well as with respect to its destination)

(f) the variation of the precise position of individuals within their socio-economic group

(g) the different criteria defining the groups.

It is therefore necessary to use a **probabilistic** approach in order to assess the expected choice behaviour, based on the assumption of individual (random) utility maximization. For this purpose the random utility theory appears to be well suitable. The general features shared by all random utility models are:

- the existence of a discrete set of choice items (alternatives);
- a partition of the population (the set of actors) into homogeneous subgroups, each having the same choice set and the same characteristics;
- the existence of an individual (random) utility function which has to be maximized over the choice set by each actor;
- average utility is made up by the expected utility of all attributes characterizing a certain choice item;
- each utility function is composed of a deterministic component and a random component, so that the total utility of an actor n with regard to a choice item i can be written as:

$$U^*_{ni} = V(z_{ni}) + \xi(z_{ni}) \tag{4.7}$$

with

U^*_{ni} : total (random) utility of choice item i (i = 1,..., I) for actor n (n = 1,..., N);

$V(z_{ni})$: the deterministic part of the utility function of actor n, being determined by the vector of attributes of alternative i, z_{ni};

$\xi(z_{ni})$: the random part of the utility function of actor n, representing individual utility differences emerging from taste variations, individual measurement errors, effects of missing data, misspecifications, etc.

Then the probability P_{ni} that actor n selects an alternative i is defined as:

$$P_{ni} = \text{Prob}\{[V(z_{ni})+\xi(z_{ni})] \geq [V(z_{ni'})+\xi(z_{ni'})]\} \qquad i' = 1,...I; \quad i' \neq i \qquad i = 1,...I; \quad n = 1,...,N \tag{4.8}$$

Expression (4.8) corresponds to the random utility maximization, characterizing the family of **Discrete Choice Models** (DCMs).

Clearly, the choice probability defined in expression (4.8) depends on the distribution of the random term $\xi(z_{ni})$. A broad family of DCMs may arise based on assumptions "on the type of distribution of the error terms (negative exponential distribution, extreme value distribution and normal distribution) and assumptions on the error terms ((in)dependently/(not)identically or general variance-covariance structure and taste variations" (Timmermans and Borgers, 1985, p. 24). Consequently, according to different specifications related to the above mentioned hypothesis concerning $\xi(z_{ni})$, we may subdivide DCMs into:

- models with Independent Identically Distributed (IID) error terms; examples are Multinomial Logit (MNL) models and some elimination by aspects models;
- closed-form models without IID error terms; examples are nested logit models, general extreme value models, and prominence theory of choice models;
- multinomial probit models.

In the seventies, one specific class of disaggregate DCMs has acquired an important position in the literature on disaggregate spatial interaction and activity analysis, viz. the MNL model. One of the major weaknesses of the frequently applied MNL model is its IID hypothesis, theoretically equivalent to the "Independence from Irrelevant Alternatives" (IIA) property (Domencich and McFadden, 1975).

In particular the IIA axiom states that the ratio of the choice probabilities of any two alternatives depends only on the ratio of their deterministic utility components and is unaffected by the existence or emergence of additional alternatives. This axiomatic idea was already put forward by Arrow (1951) in a particular social choice context, by Luce and Raiffa (1957) in decision theory, and by Luce (1959) in psychology. Consequently, on the one hand, the IIA axiom permits the introduction and/or elimination of alternatives in the choice set without a re-estimation of the parameters, but on the other hand the IIA property is considered a very restrictive assumption, in particular when the individual is facing (almost) similar alternatives (see, among others, Wrigley, 1982).

In order to overcome the IIA assumption adjusted specifications of the above mentioned MNL model have been formulated, such as the dogit model (see Gaudry and Dagenais, 1979), the decompositional multi-attribute preference model (see Timmermans, 1984, and Timmermans and Borgers, 1985), and the nested logit model as a special case of the generalized extreme value model (see Ben-Akiva and Lerman, 1979, McFadden, 1979, and for a systematic review Fischer and Nijkamp, 1985a, 1985b). Especially the nested logit model has received a strong theoretical and empirical position, as it can readily handle correlated random components of utility and hence embodies more general properties of cross-substitution than the MNL model without sacrifice of computational tractability (see also McFadden, 1978).

The most general, least restrictive of the various DCMs is the **multinomial probit model** (see Daganzo, 1979). This model allows the random components of utility of choice items to be correlated and to have unequal variances, while also random taste variations across individuals are permitted. The computational problems of multinomial probit models are however not easy to solve (see, e.g., Bunch, 1989), though the Clark approximation to reduce the calibration problem to one of sequential univariate integration (see, e.g., Sheffi and Daganzo, 1982) as well as numerical integration (see

Bunch and Kitamura, 1989), may be helpful in this respect.

Consequently, among all these DCMs, the MNL model seems the most attractive one due to its easy computational tractability. More specifically MNL models are derived from the hypothesis of IID random disturbances $\xi(z_{ni})$, distributed according to the extreme value distribution which is the following (see McFadden, 1974, 1978):

$$\text{Prob}[\xi(z_{ni}) < \xi] \leq \exp(-\exp(-\xi)) \tag{4.9}$$

Under the assumptions (4.8) and (4.9), and by assuming that $V(z_{ni})$ in (4.7) describes the average behaviour of the population (see, for example, Leonardi, 1985a), i.e.:

$$V(z_{ni}) = V(z_i) = V_i \qquad \begin{array}{l} i = 1,\dots,I \\ n = 1,\dots,N \end{array} \tag{4.10}$$

the final formulation for the probability of choosing the ith alternative is the following:

$$P_i = \exp V_i \, (\sum_k \exp V_k)^{-1} \qquad i,k \in I \tag{4.11}$$

Some authors (for example, see Fischer and Nijkamp, 1985a, 1985b, Longley, 1984, and Timmermans, 1984) start from condition (4.7) without considering expressing (4.10). It turns out that in this case the final formulation of the probability of choice is the following:

$$P_{ni} = \exp V_{ni} \, (\sum_k V_{nk})^{-1} \qquad i,k \in I \quad n \in N \tag{4.12}$$

It can easily be shown that expressions (4.11) and (4.12) are equivalent; in fact, under the hypothesis that the deterministic part V_{ni} of the utility function U^*_{ni} defined in (4.7) is linear in the attributes of the alternatives as well as of the individuals (as is usually assumed), the component related to the individual attributes vanishes in expression (4.12) so that:

$$P_{ni} = P_i \tag{4.13}$$

and hence expression (4.12) is reduced to expression (4.11).

Expression (4.11) shows the drawback inherent in all MNL models, that is, no constraint related to individual attributes (for example, budget constraints) is explicitly included. To avoid the above mentioned problem, one usually partitions the population of actors into relatively homogeneous population segments with more or less similar socioeconomic characteristics, but this is evidently not a completely satisfactory and rigorous method.

In order to overcome this drawback, attention has recently been paid to the integration of MNL models with other methodologies able to identify a threshold for selecting 'efficient' alternatives, such as the so-called stochastic dominance approach (see Reggiani and Stefani, 1986, 1989).

Finally, it should be noted that it has also been shown (see Reggiani and Stefani, 1986) that DCMs are consistent with the framework of decision-making under risk, particularly with the well-known maximum utility decision rule. In the next section we will investigate the formal connection between SIMs emerging from the entropy-maximizing approach and DCMs emerging from micro-economic theory, in a static framework. These relationships will be further elaborated in a dynamic context (see Chapter 5).

4.2.3.2 Analogies between spatial interaction models and discrete choice models

In the previous analysis it has been shown that SIMs have a macro foundation, based on entropy maximization (see Chapters 2 and 3), while DCMs have a micro foundation, based on individual utility maximization (see Section 4.2.3.1).

The first interesting formal analogy between SIMs and DCMs can be seen by transforming the standard solution (2.15) of the entropy maximization problem (2.8-2.9) as follows:

$$P_{ij} = \frac{T_{ij}}{O_i} = A_i B_j D_j \exp(-\beta c_{ij})$$

$$= \frac{B_j D_j \exp(-\beta c_{ij})}{\sum_j B_j D_j \exp(-\beta c_{ij})} \qquad (4.14)$$

where P_{ij} represents the probability of a destination choice from i to j.

It is clear that (4.14) also represents the formulation of an MNL model defined in (4.11) where, according to Wegener et al. (1985), $-c_{ij}$ can be considered to be equal to the deterministic part of the individual random utility function defined in (4.7) and (4.10), and $B_j D_j$ can be interpreted as a weighting factor reflecting the attractiveness of a point of destination j.

In particular, because B_j and A_i are interrelated (see (2.16) and (2.17)), the resulting P_{ij} will not only depend on the characteristics of destination j and origin i, but also on the characteristics of other origins. This does not happen for instance in the case of a singly production-constrained SIM, because in that case the term B_j is equal to 1 in (4.14).

Therefore, it can be argued that while the singly-constrained SIM can be considered equivalent to an MNL model, the doubly-constrained SIM represents a more general form than the MNL model owing to the more comprehensive factor B_j. Moreover, if factor B_j is regarded as a measure of accessibility from j to other destinations K (where K represents for instance a set of subregions in j), model (4.14) incorporates a **sequential choice structure** and can be compared to a nested MNL model (see, for a proof, Reggiani, 1990), thus encompassing the IIA axiom. In this context it is interesting to mention also Anas (1983) who demonstrates the equivalence between doubly-constrained SIMs and logit models of joint origin-destination choice, as well as the discussion offered by Fotheringham (1986) on competing destination models incorporating a hierarchical destination choice.

The same observation holds obviously for the Alonso model illustrated in Section 2.3. Starting from equation (2.26) we obtain the final probability P_{ij} that a trip starting in i will end up in j as follows:

$$P_{ij} = \frac{T_{ij}}{S_i W_i^{\sigma_i}} = \frac{Y_j Z_j^{\tau_j - 1} \exp(-\beta c_{ij})}{\sum_j Y_j Z_j^{\tau_j - 1} \exp(-\beta c_{ij})} \tag{4.15}$$

which, by using the equivalence:

$$\sum_i T_{ij} = Y_j Z_j^{\tau_j} = D_j \tag{4.16}$$

leads at the end to the following result:

$$P_{ij} = \frac{D_j Z_j^{-1} \exp(-\beta c_{ij})}{\sum_j D_j Z_j^{-1} \exp(-\beta c_{ij})} \tag{4.17}$$

It is clear that (4.15) represents a more general form than (4.14) owing to the response rates σ_i and τ_j, included in (4.15) and depending on the rest of the system. Thus (4.17) may also be interpreted as a more general nested MNL model, where the term B_j is replaced by the term Z_j^{-1}, but essentially with the same meaning as indicated previously.

The mathematical equivalence of SIMs and MNL models has also been demonstrated by Anas (1983). He argues that entropy and gravity models are not inherently less behavioural than stochastic utility models of discrete choice and MNL models in particular. A similar 'economic' interpretation has also been advocated by Leonardi (1985a), who starts from a micro theory (the maximization of stochastic utility) and - by introducing statistical techniques of aggregation (based on the so-called asymptotic methods) - then obtains the MNL model. Besides the above mentioned authors, many others have shown the formal correspondence between entropy theory and random utility theory (for a broad systematic review see Nijkamp and Reggiani, 1988a, 1989a). It is interesting to mention also Mattsson (1983) who has shown the equivalence between an aggregate expected utility approach and entropy maximization. And finally, Roy and Lesse (1985) show that entropy models (in terms of probability) with entropy constraints on aggregated events can be associated with nested MNL models.

4.2.4 Concluding remarks

The fundamental result from this section, in conjunction with the findings from Chapters 2 and 3, is clearly that the **entropy models can be reinterpreted in a behavioural context with an economic meaning**, suggesting a unified theoretical framework. Thus the equivalent SIM, being consistent with micro-economic theory at both deterministic and stochastic levels, can be considered as an **aggregate model of human behaviour**.

It turns out that "the same model without any aggregation error may be derived in disaggregate form from both entropy and utility maximization and can therefore be estimated" (Batten and Boyce, 1986, p. 378). However, various questions are still left unanswered, especially if we consider the emerging relationships between SIMs and DCMs. Examples are:

- The MNL model is only a specific case of DCM; if there is a correspondence between MNL models and SIMs, there might perhaps also exist a direct compatibility between other categories of DCMs (e.g., probit, general extreme value models) and entropy type of models, but so far no unambiguous solution for this complicated analytical problem has been found.
- Consequently, the limits and axioms inherent in DCMs (e.g., the IIA assumption) also apply to the singly-constrained SIMs. In this context the doubly-constrained SIM with a sequential choice process - and more generally the Alonso model - may potentially be more interesting, since they can be linked to a nested MNL model (thus overcoming the IIA axiom).
- A final consequence emerging from the equivalence between MNL and singly-constrained SIMs is that SIMs implicitly assume that the utility functions underlying the mean utility $-c_{ij}$ are linear in the attributes of the alternatives as well as of the individuals, while so far essentially no hypothesis has been formulated about the utility (or cost) function in SIMs.

However, despite all these considerations which need more attention in future research, the formal equivalence between SIMs and MNLs offers a **new theoretical strength to SIMs**, in the sense that SIMs can be considered as aggregate models of human behaviour so that also empirical applications using SIMs now have a deeper theoretical justification. This will also be the subject of the next section.

4.3 Spatial Interaction Behaviour and Individual Choice Theory: An Application

4.3.1 Introduction

In the previous section it has been shown that DCMs and SIMs offer two largely equivalent - or at least mutually compatible - views on the same spatial choice or allocation problem; whilst the source of the former approach seems more 'behavioural' and the source of the latter more 'informational', the final result is the same. Historically the different use of data (micro versus macro) within these two paradigms obscured the (large degree of) similarity between the two methods: if we would use disaggregated data in a SIM, we would directly reach a synthesis with DCMs (see Anas, 1983).

Consequently it is now interesting to show how SIMs (or MNL models) can be developed and used for analyzing a transport system by focussing on the mobility pattern of trip-makers. Travel decisions are discrete in nature, since choices are made among discrete sets of alternatives, such as routes and modes of travel, destinations, etc. Thus, DCMs or SIMs seem well suitable for representing and analyzing this type of choice problems.

In particular the present study pays attention to the utility function in an MNL model, applied to a modal split analysis, by incorporating a dynamic function related to congestion phenomena; the underlying hypothesis is that congestion may generate new steady states (in the course of time) in the user's distribution converging towards a final stable solution.

For this purpose a network transport problem with one origin and one destination point has been examined (notably, the Bergamo-Milan corridor) by using the Census Data 1981, which are disaggregated into four main socio-economic categories of users. In order to examine the congestion phenomena two daily time intervals of use (in the morning hours) will be considered.

4.3.2 The model

An MNL model related to a modal choice problem can be expressed as follows (see for the theoretical background Section 4.2.3).

$$P_{hm} = \frac{\exp(V_{hm})}{\sum_{m'} \exp(V_{hm'})} \qquad m,m' = 1,...,M \qquad h = 1,...,H \tag{4.18}$$

where P_{hm} represents the choice probabilities of mode m for individual h. V_{hm} is the utility function which, as usual, is assumed to be linear in the measurable attributes $\mathbf{A}_m$ of the alternatives m:

$$V_{hm} = \alpha_h A_m \qquad h = 1,...,H \qquad m = 1,...,M \tag{4.19}$$

where:

$$\alpha_h = [\alpha_{h1}, ...\alpha_{hj}, ...\alpha_{hn}] \tag{4.20}$$

$$A_m = [A_{m1}, ...A_{mj}, ...A_{mn}] \tag{4.21}$$

and n is the number of the attributes related to mode m. In our specific case h is represented by means of four socio-economic categories H_i (i = 1,...,4). Consequently in (4.21) $\mathbf{A}_m$ represents the vector of all attributes related to mode m (such as time, cost, frequency, waiting time, etc.), while in (4.20) α_h represents the vector of the relevant parameters - for each socio-economic category -associated with the attributes $\mathbf{A}_m$ of mode m.

A further step in our analysis is the introduction in (4.19) of a dynamic function $B_{hm}^{(t)}$ indicating the congestion effect, at time t, as follows:

$$V_{hm}^{(t+1)} = \alpha_h A_m + \beta_h B_{hm}^{(t)} \qquad m = 1,...,M \qquad h = 1,...,H \tag{4.22}$$

where β_h is essentially an elasticity coefficient (related to income class h) of the dynamic congestion $B_{hm}^{(t)}$. In (4.22) $B_{hm}^{(t)}$ is defined as (see also Figure 4.1):

$$B_{hm}^{(t)} = \begin{cases} 0 & 0 \le D_m^{(t)} \le K_m^{(s)} \\ \log \dfrac{D_m^{(t)}}{K_m^{(s)}} & K_m^{(s)} \le D_m^{(t)} \le K_m \\ \log \dfrac{K_m}{K_m^{(s)}} & K_m < D_m^{(t)} \end{cases} \qquad \begin{array}{l} m = 1,\ldots,M \\ \\ h = 1,\ldots,H \end{array} \tag{4.23}$$

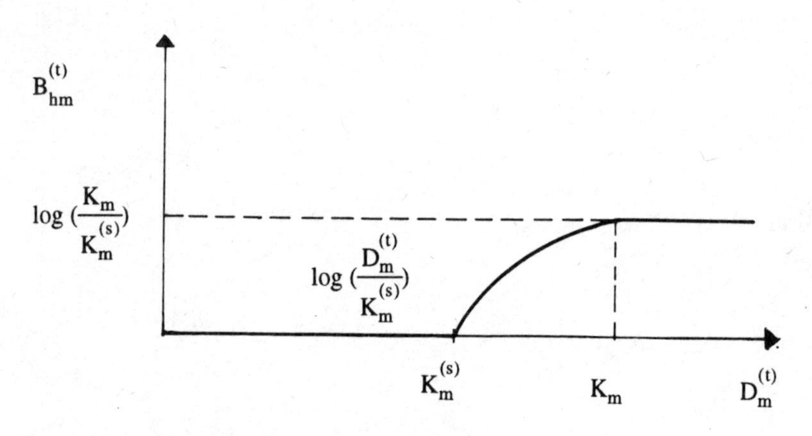

Figure 4.1 The Dynamic Congestion Function

In (4.23) $D_m^{(t)}$ represents the number of people using mode m, whilst K_m is the global capacity of mode m, which can be expressed as the sum of the number of seats ($K_m^{(s)}$) plus the capacity for standing passengers ($K_m^{(f)}$):

$$K_m = K_m^{(s)} + K_m^{(f)} \tag{4.24}$$

In particular, $K_m^{(s)}$ represents a threshold beyond which it is assumed that congestion follows a logarithmic evolution (since it seems plausible that dynamic congestion follows a concave shape, with decreasing marginal utility) up to the limit $D_m^{(t)} = K_m$.

In conclusion the introduction of $B_{hm}^{(t)}$ allows to model the variations, in the utility function, of an MNL model, due to congestion effects and consequently the variations in the users' distribution toward a final stable configuration.

4.3.3 The data

The data used in this study are mainly derived from the 1981 Census. These are subdivided according to income class, travel time, mode of transport, and time intervals during the day.

As regards the attributes of the modes, besides the travel time (included in the above mentioned Census), the attributes 'frequency', 'capacity' and 'cost' (information provided by the Italian Ferrovie dello Stato e Società Autostrade) are taken into considation as well. In particular, in this model the attribute 'frequency x capacity' has been used to indicate a kind of 'availability' of the mode.

The income classes have been aggregated into four main categories: professionals (H_1), managers (H_2), employees (H_3), and students (H_4). The time intervals considered are two, according to the Census data, viz., 7.15 a.m. - 8.14 a.m. and 8.15 a.m. - 9.14 a.m.

The number of modes is six: train (M_1), bus (M_2), car with driver only (M_3), car with driver and passengers (M_4), motorcycle (M_5), other forms of transport (e.g., taxi) (M_6). There are four attributes which are considered relevant: time (A_1), availability (A_2), price (A_3) and dynamic congestion (A_4).

The parameter estimation of function (4.22) is carried out by means of the GLIM programme (cf. Baker and Nelder, 1978) (PC version) where in a first stage no dynamic congestion has been included (i.e.,the factor $B_{hm}^{(t)}$). The dynamic factor $B_{hm}^{(t)}$ is then introduced in (4.22) and utilized for the estimation of the final user's distribution (see also Annex 4A). The results are presented in the next section.

4.3.4 Results and concluding remarks

In Tables 4.1 and 4.2 the partial utilities related to the various forms of transport are shown for each time interval. The emerging results are clearly significant: firstly, as we could easily suppose, the utility related to the attribute 'availability' (V_2) is always positive, for each income class and each time interval.

Secondly, the utility related to the attribute price (V_3) does not seem very relevant in comparison to the congestion effect (V_4), for all income classes. According to the current literature (see, e.g., Ahsan, 1982) it seems that the travel conditions are the most important factors in the mode choice.

In particular from Tables 4.1 and 4.2 the importance of the congestion effect, in the two times intervals, is clear, at least for the first three categories (professionals, managers and employees). Obviously for the students the negative effect of congestion is not so high as for the other users. The significance of the congestion factor can also be verified in Tables 4.3 and 4.4 (related to parameter estimates). In these tables it is also evident that congestion is not a significant variable for students, whereas this is certainly so for the other three categories.

If we analyze more thorougly the total utility function for each income class, we can conduct other interesting investigations on the behaviour of the users. In particular we can see that the final total utility has high positive values in the first interval time for all categories. In the second time interval the total utility has less high positive values or is negative; we can thus infer the conclusion that these passengers prefer to arrive in Milan early in the morning. In fact we observe that professionals (H_1) show a clear deterrence in travelling in the second time interval; with regard to managers (H_2) and employees (H_3) we see a decreasing utility in the second hour; finally, concerning the students (H_4) we can always notice their positive utility in travelling, since they are not obliged to attend the morning lectures.

Then from Tables 4.1 and 4.2 we can also derive the final choice of the mode of transport. For example, we can see that professionals (H_1) tend to choose the car. The managers (H_2) appear to prefer the train, and the car too; the employees (H_3) attach a high preference to train and bus, while students (H_4) always prefer the cheapest train.

The final distribution (obtained from the algorithm illustrated in Annex 4A) can also be compared to the observed distribution (see Tables 4.5 and 4.6). The good results in this respect confirm the fact that this new utility function, also incorporating a dynamic congestion phenomenon, provides an appropriate tool for simulating and/or forecasting user's behaviour, in the choice of mode of transport under a situation of congestion.

Table 4.1 Partial and Total Utilities Related to the Attributes of the Different Modes of Transport in the First Interval Time (7.15 a.m. - 8.14 a.m.)

H_1	V_1	V_2	V_3	V_4	V_{tot}
M_1	0.95836	.01325	-0.06513	-5.257	1.6497
M_2	0.95833	.34069	-0.18007	-2.934	1.1853
M_3	0.54763	.70683	-0.38313	0	3.8713
M_4	0.54761	.75720	-0.07663	0	2.2282
M_5	0.47920	.24320	-0.30650	0	0.4159
M_6	0.4792	0.09488	-1.91565	0	-1.3416
H_2					
M_1	7.330	6.7761	0.01945	-5.400	8.725
M_2	7.330	3.7645	0.05378	-3.014	8.134
M_3	4.188	4.1771	0.11442	0	8.480
M_4	4.188	1.9801	0.02288	0	6.191
M_5	3.665	0.2741	0.09154	0	4.030
M_6	3.665	0.1096	0.57210	0	4.344
H_3					
M_1	3.838	2.82432	-0.05567	-2.590	4.0164
M_2	3.838	1.56907	-0.18905	-1.445	3.7723
M_3	2.193	1.74104	-0.52515	0	3.4088
M_4	2.193	0.82533	-0.10503	0	2.9132
M_5	1.919	0.11423	-0.42012	0	1.6129
M_6	1.919	0.04456	-2.62573	0	-0.6624
H_4					
M_1	3.205	3.33051	-0.02164	-0.3721	6.1417
M_2	3.205	1.85028	-0.09060	-0.2077	4.7570
M_3	1.831	2.05307	-0.25168	0	3.6328
M_4	1.831	0.97325	-0.05034	0	2.7543
M_5	1.602	0.13470	-0.20134	0	1.5358
M_6	1.602	0.05255	-1.25839	0	0.3966

Legend:

H_1 = professionals
H_2 = managers
H_3 = employees
H_4 = students

M_1 = train
M_2 = bus
M_3 = car with driver only
M_4 = car with driver and passengers
M_5 = motorcycle
M_6 = other forms of transport

V_1 = utility related to time
V_2 = utility related to availability
V_3 = utility related to price
V_4 = utility related to congestion

Table 4.2 Partial and Total Utilities Related to the Attributes of the Different Modes Transport in the Second Interval Time (8.15 a.m. - 9.14 a.m.)

H_1	V_1	V_2	V_3	V_4	V_{tot}
M_1	-3.758	9.65940	-0.08962	-6.284	-0.4715
M_2	-3.758	2.92709	-0.24778	0	-1.0784
M_3	-2.192	4.00133	-0.52718	0	1.2822
M_4	-2.505	0.72299	-0.10544	0	-1.8876
M_5	-2.192	0.12294	-0.42175	0	-2.4908
M_6	-2.192	0.07610	-2.63591	0	-4.7518
H_2					
M_1	0.6004	8.75498	-0.1201	-5.334	3.9014
M_2	0.6004	2.65303	-0.3320	0	2.9214
M_3	0.3502	3.62669	-0.7063	0	3.2706
M_4	0.4002	0.65530	-0.1413	0	0.9143
M_5	0.3502	0.11143	-0.5651	0	-0.1034
M_6	0.3502	0.06898	-3.5316	0	-3.1124
H_3					
M_1	1.3883	4.71267	-0.02263	-2.650	3.4280
M_2	1.3883	1.42808	-0.07687	0	2.7395
M_3	0.8099	1.95219	-0.21354	0	2.5485
M_4	0.9255	0.35274	-0.04271	0	1.2356
M_5	0.8099	0.05998	-0.17083	0	0.6990
M_6	0.8099	0.03713	-1.06768	0	-0.2207
H_4					
M_1	3.320	2.81846	-0.03197	-0.2472	5.8596
M_2	3.320	0.85408	-0.13383	0	4.0406
M_3	1.937	1.16753	-0.37174	0	2.7326
M_4	2.214	0.21096	-0.07435	0	2.3502
M_5	1.937	0.03587	-0.29739	0	1.6753
M_6	1.937	0.02221	-1.85871	0	0.1004

Legend:

H_1 = professionals
H_2 = managers
H_3 = employees
H_4 = students

M_1 = train
M_2 = bus
M_3 = car with driver only
M_4 = car with driver and passengers
M_5 = motorcycle
M_6 = other forms of transport

V_1 = utility related to time
V_2 = utility related to availability
V_3 = utility related to price
V_4 = utility related to congestion

Table 4.3 Parameter Estimates Related to the Attributes of the Different Modes of Transport in the First Time Interval (7.15 a.m. - 8.14 a.m.). (Standard errors are given in brackets).

H_1	Time	Availability	Price	Congestion
estimates	0.01369	1.336	-0.03831	-15.92
	(0.01419)	(0.06526)	(0.01144)	(1.629)
H_2				
estimates	0.1047	1.506	0.01144	-16.36
	(0.006044)	(0.05665)	(0.006597)	(0.8266)
H_3				
estimates	0.05482	0.6276	-0.05251	-7.844
	(0.002133)	(0.01993)	(0.003516)	(0.330)
H_4				
estimates	0.04579	0.7401	-0.02517	-1.127
	(0.009473)	(0.1030)	(0.01381)	(1.495)

Table 4.4 Parameter Estimates Related to the Attributes of the Different Modes of Transport in the First Time Interval (8.15 a.m. - 9.14 a.m.). (Standard errors are given in brackets).

H_1	Time	Availability	Price	Congestion
estimates	-0.06263	2.927	-0.05272	-31.17
	(0.01247)	(0.2680)	(0.03210)	(3.631)
H_2				
estimates	0.01001	2.653	-0.07063	-26.46
	(0.003275)	(0.1059)	(0.01538)	(1.380)
H_3				
estimates	0.02314	1.428	-0.02135	-13.15
	(0.0040)	(0.09986)	(0.006586)	(1.212)
H_4				
estimates	0.05534	0.8541	-0.03717	-1.227
	(0.005413)	(0.1412)	(0.01044)	(1.679)

Table 4.5 Theoretical and Observed Distributions per Mode in the First Time Interval (7.15 a.m. - 8.14 a.m.).

H_1	observed travellers	predicted travellers
M_1	52	50
M_2	29	32
M_3	459	465
M_4	105	90
M_5	5	15
M_6	5	3
H_2		
M_1	964	965
M_2	537	535
M_3	761	756
M_4	64	77
M_5	17	9
M_6	10	12
H_3		
M_1	2892	2875
M_2	2223	2252
M_3	1495	1566
M_4	1117	954
M_5	155	260
M_6	53	27
H_4		
M_1	746	746
M_2	186	187
M_3	59	61
M_4	29	25
M5	5	7
M_6	3	2

Legend:

H_1 = professionals
H_2 = managers
H_3 = employees
H_4 = students

M_1 = train
M_2 = bus
M_3 = car with driver only
M_4 = car with driver and passengers
M_5 = motorcycle
M_6 = other forms of transport

Table 4.6 Theoretical and Observed Distributions per Mode in the Second Time Interval (8.15 a.m. - 9.14 a.m.).

H_1	observed travellers	predicted travellers
M_1	31	31
M_2	16	17
M_3	179	179
M_4	12	8
M_5	0	4
M_6	1	0
H_2		
M_1	1701	1701
M_2	633	638
M_3	905	905
M_4	113	86
M_5	6	31
M_6	5	2
H_3		
M_1	499	499
M_2	249	251
M_3	207	207
M_4	64	56
M_5	25	33
M_6	14	13
H_4		
M_1	1439	1439
M_2	231	234
M_3	63	63
M_4	55	43
M5	11	22
M_6	6	5

Legend:

H_1 = professionals
H_2 = managers
H_3 = employees
H_4 = students

M_1 = train
M_2 = bus
M_3 = car with driver only
M_4 = car with driver and passengers
M_5 = motorcycle
M_6 = other forms of transport

4.4 Conclusions

In the first part of the present chapter the attention has been devoted to relationships between SIMs and utility maximizing behaviour at the micro level. In particular it has been shown that SIMs result in a consistent way from the maximization of both deterministic and random (individual) utility. This latter result seems more interesting since it represents a more realistic behaviour, where population is heterogeneous. Hence it also appeared that a formal equivalence exists between SIMs and DCMs emerging from random utility theory. In particular a singly-constrained SIM can be associated with an MNL model, while a doubly-constrained SIM with a sequential choice process - and more generally an Alonso model - appears to comply with a nested MNL model.

Furthermore the strong equivalence between SIMs and MNL models is also reflected in the coefficient estimates; it can be demonstrated that "the behavioural content of models estimated by either approach is entirely determined by the model specification and data aggregation belief of the analyst, and not by any inherent structural property of the models themselves" (Anas, 1983, p. 13).

The main result is therefore the confirmation of a behavioural economic interpretation of entropy, which already emerged in the previous chapters. This sheds new light on the multiple aspects of entropy and its ability of capturing the essence of a spatial system; it is therefore also worthwhile to examine the 'entropy-utility' concept and its connection with SIMs also in a dynamic context (see in particular Chapter 5).

A further result is the theoretical strength of doubly-constrained SIMs embedding a sequential choice process (and hence of the Alonso model), since these models - being linked to a nested MNL model - are not hampered by IIA assumptions.

Consequently, it is evident that the two seemingly alternative approaches, viz., macro entropy and micro discrete choice theory, are two equivalent views on the same problem; and either approach is more general than it seems at first glance. From this perspective it is easy to identify new future research directions; on the one hand there is a need for empirical investigation of the potential of these two approaches with particular attention for the specification of the underlying utility function, and on the other hand more attention should be given to the exploration of the connection between the above mentioned macro and micro approaches in a dynamic context. The latter topic will be dealt with in Chapter 5 and subsequent chapters, while the first topic has already been dealt with in the second part of the present chapter by considering an application of a SIM to a modal split choice problem.

This illustration shows the relevance of applying SIMs in a geographical context; thus different specifications of SIMs (with different variables in the utility function) have to be used according to the underlying behavioural processes in different geographical settings. In this context it should be noted that a quasi-dynamic utility function may also be used by introducing a dynamic congestion factor which may generate changes in the users distribution (leading to a new stable distribution).

Finally, the need for a new form of analysis based on dynamic and uncertain elements became evident from this chapter. Consequently, after the discussion and development of the great variety of features related to static SIMs (and in particular their connections with entropy theory and economic theory) we will now develop new departures related to dynamic spatial interaction modelling. This will be the subject of the second part of the present volume.

Annex 4A

An Algorithm for Modal Split Choice with Congestion

In this annex we will describe an iterative procedure in order to take into account the changes in the user's distribution due to congestion factors. For this purpose we will use an algorithm which introduces in the utility function of a logit distribution (estimated by means of the standard GLIM programme) the dynamic function of congestion $B_{hm}^{(t)}$ as defined in (4.22). This congestion effect $B_{hm}^{(t)}$ will produce a new configuration in the users' distribution which will be considered stable when the convergence condition of step 3 will be satisfied.

Step 0: Set t = 0 and $D_m^{(o)} = 0$.

Step 1: Estimate the utility function $V_{hm}^{(t+1)}$ from (4.22).

Step 2: Calculate the user's distribution by means of the logit formula:

$$D_m^{(t+1)} = \sum_{h=1}^{H} D_{hm}^{(t+1)} = \sum_{h=1}^{H} S_h P_{hm}^{(t+1)} = = \sum_{h=1}^{H} S_h \frac{\exp[V_{hm}^{(t+1)}]}{\sum_{m'=1}^{M} \exp[V_{hm'}^{(t+1)}]} = \sum_{h=1}^{H} S_h \frac{\exp[\alpha_h A_m + \beta_h B_{hm}^{(t)}]}{\sum_{m'=1}^{M} \exp[\alpha_h A_{m'} + \beta_h B_{hm'}^{(t)}]} \quad (4A.1)$$

where S_h = total number of persons belonging to category h.

Step 3: Check the convergence condition:

$$\text{if: } |D_m^{(t+1)} - D_m^{(t)}| < \delta \qquad \delta > 0 \text{ fixed} \quad (4A.2)$$

then STOP.

Step 4: Calculate the congestion factor.

Set:

$$t = t+1 \tag{4A.3}$$

Calculate the congestion effect $B_{hm}^{(t)}$ from (4.23).

Return then to step 1.

This recursive procedure appears to lead to efficient results.

variation of income [illegible] probability [illegible] by the stochastic logit form, which [illegible] moving toward j after [illegible] from i. Obviously [illegible] some similarity [illegible] Tomlin (1979).

5.3.3 Concluding [illegible]

In this [illegible] SIM can [illegible] structure which is the result of [illegible] and the (stochastic) destination [illegible] synthesis [illegible]

Another [illegible] equation (5.19) is formally [illegible] and explanatory [illegible] related to the maximization of entropy [illegible] It should also be noted that the use of [illegible] stochastic context is interesting: multi-objective [illegible] (see, for instance Nijkamp, 1977, 1979b, [illegible] stochasticity (see, for example, [illegible]

In the next section [illegible] analyzing the possibility of catastrophic [illegible]

5.4 Spatial Interaction [illegible] Analyzed in the Framework of Stochastic Optimal Control

5.4.1 The model

In this section we will pay attention to a production-constrained SIM by extending [illegible] with the production-constrained [illegible] and Hansen [illegible] with respect to the [illegible] the previous section, we will consider a migration context [illegible] the following [illegible]

PART B

DYNAMIC MODELS OF SPATIAL INTERACTION

CHAPTER 5

DYNAMIC AND STOCHASTIC SPATIAL INTERACTION MODELS

5.1 Prologue

In the first part of this study the attention has been focussed on static SIMs. The second part of the present study will be devoted to the analysis of dynamic modelling, with particular reference to dynamic SIMs.

As noticed before, spatial dynamics has become an issue of great interest in the past decade, mostly for its capability of modelling the evolution of urban and regional systems (see Beckmann and Puu, 1985, 1990, Bertuglia et al., 1987, Griffith and Lea, 1983, Hauer et al., 1989, Isard and Liossatos, 1979, and Wilson, 1985). Moreover, the emerging possibility of modelling stochastic and chaotic elements in spatial dynamics is generating many interesting studies about new mathematical and computational forms inherent in dynamic processes.

Consequently, the second part of this book will introduce various recent approaches and methodologies for dynamic systems analysis which may offer new and interesting contributions to spatial interaction modelling (such as stochastic optimal control analysis, ecologically-based theories, catastrophe and bifurcation theory, chaos theory).

In particular the first part of the present chapter is a straightforward generalization and extension of Chapters 2 and 4, since it deals with the analytical derivation of a SIM (and hence its correspondence with an MNL model) in a dynamic (and also stochastic) context.

More specifically, an optimal control model based on a cumulative entropy function will be developed and a formal correspondence between macro dynamic models for spatial interaction analysis and micro choice models will be shown. Next the possibility of introducing both stochastic and multiobjective elements in such a dynamic framework of SIMs will be explored. In this framework a stochastic dynamic multiobjective SIM that results in a solution compatible with an MNL model will be developed. Then, we will analyze another type of stochastic dynamic multiobjective SIM which can give rise, under certain conditions, to catastrophe behaviour of the Wilson type. The interesting result emerging from these two research issues is a stochastic movement stemming from deterministic equations which are formally similar to the usual SIMs analyzed so far. Finally, some general remarks concerning our results will be presented.

5.2 Spatial Interaction Models Analyzed by Means of Optimal Control

5.2.1 Introduction

Various modes of analogy between SIMs and DCMs in a dynamic framework may be considered. In general, an analysis of dynamic SIMs can be distinguished according to the role of actors as follows:

a) analysis of the demand side of SIMs
b) analysis of the supply side of SIMs
c) integration of supply and demand in SIMs

In the past years interesting approaches have emerged in developing these three types of analysis.

In particular -with reference to type a)- it is worth recalling the contribution by Coelho (1977) who formulated a dynamic version of entropy for the well-known Lakshmanan-Hansen shopping model. Furthermore, adjusted methods, based on Bayesian approaches to a multi-period entropy formulation, have been put forward, among others, by Batty and March (1976) (see also Chapter 2). Another interesting approach is the introduction of congestion effects in the cost constraint leading to alternative expressions for an SIM (see for instance Annex B in Nijkamp and Reggiani, 1989a).

As regards type b) - the analysis of the supply side of SIMs - it is necessary to mention here the fundamental contribution by Harris and Wilson (1978) which has often been referred to (and elaborated on) in a large number of other interesting analyses (see, e.g., Birkin and Wilson, 1989, Harris et al., 1982, Leonardi, 1981a, 1981b, Lombardo and Rabino, 1983, Nijkamp and Reggiani, 1988c, Oppenheim, 1986, and Phiri, 1980). Harris and Wilson's basic dynamic model is related to a spatial retail system, but can easily be formulated in a more general setting. The model reads as follows:

$$\dot{W}_j = \nu (D_j - K_j W_j) \tag{5.1}$$

where W_j is the level of retail facilities in zone j, $\dot{W}_j$ the rate of change, D_j the revenue associated with j, v the rate of response of the system (when it is not in equilibrium), and K_j the unit cost of supply.

Model (5.1) is essentially a logistic equation; it is a member of a more general class of growth models which can be represented for a single variable x by (see Wilson, 1981):

$$\dot{x} = \nu x^n (d - x) \tag{5.2}$$

where a dot (.) represents a time rate of growth of the variable concerned; n is a parameter and D the carrying capacity. Since model (5.2) can be rewritten in terms of the well-known family of May-models (see May, 1976), it is evident that from (5.1) and (5.2) various kinds of dynamic behaviour may emerge, including catastrophe behaviour and bifurcations (see also Chapter 6 of the present study).

Other general dynamic models based on logistic equations analyzing structural changes in space and time can be found, among others, in Allen et al. (1978) and Sonis (1984). These types of models will also be treated and unified in a SIM framework in Chapter 7.

Finally, concerning type c) -integration of supply and demand in SIMs- so far only a few attempts have been made (see Leonardi, 1983), mainly owing to the complexity of the dynamic systems concerned. Consequently, in the next subsection attention will be paid to a new formulation of a dynamic entropy model which incorporates feedback effects from the supply side via a dynamic adjustment of the outflows O_i. For this purpose an optimal control model will be developed.

5.2.2 An optimal control approach

In this section an **optimal control version of a dynamic entropy model** will be presented by assuming a transport system in which all variables T_{ij} are time-dependent. The choice of control variables and state variables in an optimal control model is always somewhat arbitrary, as it is not always sure whether a control variable means a direct intervention measure or an indirect policy instrument. In our case, the control variable T_{ij} is obviously an indirect policy instrument, which refers to the possibility of influencing spatial movements T_{ij} by physical restraints (e.g., capacity limits) or by financial incentives (e.g., user charges). In a more comprehensive model, a clearer distinction between instruments, targets, state variables and irrelevant variables would have to be made (see also Tinbergen, 1956). Thus in our model the flows T_{ij} will be considered as control variables (e.g., by means of prices via taxes), while the origin variables O_i will be regarded as state variables. Their evolution is assumed to be dependent on the net push-out/pull-in effects of place i as follows (see also Nijkamp and Reggiani, 1988b):

$$\dot{O}_i = \tau_i O_i + \phi_i \left(\sum_j T_{ji} - \sum_j T_{ij}\right) \tag{5.3}$$

Equation (5.3) is essentially a direct derivation of the well-known dynamic migration model developed by Okabe (1979) and Sikdar and Karmeshu (1982):

$$\dot{M}_i = \tau_i M_i + \sum_j T_{ji} - \sum_j T_{ij} \tag{5.4}$$

In equation (5.4) τ_i is the natural growth of population M_i at the ith place. Clearly, equation (5.3) is derived from equation (5.4) by assuming that the O_i's are linearly dependent - through the parameter ϕ_i - on population P_i, i.e.,

$$O_i = \phi_i M_i \tag{5.5}$$

Then obviously, it follows:

$$\dot{O}_i = \phi_i \dot{M}_i \tag{5.6}$$

Therefore, substitution of (5.5) and (5.6) into (5.4) leads to (5.3). Since the parameter ϕ_i has the clear meaning of 'inclination to mobility', also congestion effects in the sense of negative attraction effects might be taken into account by making, for instance, ϕ_i a decreasing function of the net transport flows.

Finally, the objective function of the present spatial interaction system is assumed to be a multiperiod cumulative entropy function (see also Sonis, 1986) reflecting a general utility (or social welfare) of the system during the time horizon $\tilde{T}$ (cf. Coelho, 1977, and Wilson, 1981). This formulation makes also indirectly sense since the entropy has an economic/utility meaning, based on generalized cost minimization (see Chapter 3).

The resulting optimal control entropy model then reads as follows:

$$\max w = \int_0^{\tilde{T}} - \sum_i \sum_j T_{ij} (\ln T_{ij} - 1) dt$$

subject to

$$\begin{aligned} &\sum_j T_{ij} = O_i && \forall i \\ &\sum_i T_{ij} = D_j && \forall j \\ &\sum_i \sum_j T_{ij} c_{ij} = C && \\ &\dot{O}_i = \tau_i O_i + \phi_i \left(\sum_j T_{ji} - \sum_j T_{ij}\right) && \end{aligned} \tag{5.7}$$

It should be noted that it might be more in harmony with dynamic and multiperiod evaluation analysis to include a discount factor in the integrand of the objective function, but this would not fundamentally change the results (see for a proof Nijkamp and Reggiani, 1988b).

Then the solution to problem (5.7) is the following generalized spatial interaction model:

$$T_{ij} = A_i^* B_j^* O_i D_j \exp(-\beta c_{ij}) \tag{5.8}$$

where:

$$A_i^* = [\sum_j B_j^* D_j \exp(-\beta c_{ij})]^{-1} \tag{5.9}$$

and:

$$B_j^* = [\sum_i A_i^* O_i \exp(-\beta c_{ij})]^{-1} \tag{5.10}$$

Thus the result of this optimal control model is again a standard SIM. The proof of this interesting result is given in Annex 5A. Solution (5.8) is unique, as the objective function of problem (5.7) is concave (analogous to the static case). Furthermore, from (5.8) and (5.9) the dynamic probability of choice from i to j can easily be derived:

$$P_{ij} = \frac{T_{ij}}{O_i} = \frac{B_j^* D_j \exp(-\beta c_{ij})}{\sum_j B_j^* D_j \exp(-\beta c_{ij})} \tag{5.11}$$

Expression (5.11) is the dynamic version of expression (4.14), as in (5.11) all variables are a function of time. Furthermore, also the balancing factor B_j^* is dynamic, since it depends on the costate variable Φ_j (see Annex 5A), which might be interpreted as dynamic accessibility to zones j. It should be noted that the latter concept stems from the interpretation of the Lagrange multipliers λ_i, and μ_j in (5A.12) and (5A.13) as 'accessibility potentials' since they reflect the 'ease of flow' from the origin zones i and to the destination zones j, respectively, measured by the variation of the entropy of the system (cf. Coelho, 1977). In conclusion, solution (5.11) is formally equivalent to a generalized dynamic MNL model.

Obviously, this surprising and interesting result strongly depends on the analytical form of the additional dynamic equation in (5.7); other expressions for the evolution of the state variable do not necessarily lead to the standard result of a SIM (see again Annex 5A).

5.2.3 Concluding remarks

The main conclusion emerging from this section is that the optimal **solution of**

an optimal control problem - with a dynamic/cumulative entropy function as objective function and subject to a certain dynamic equation for the outflows O_i - **is a generalized SIM.** This result also demonstrates the formal correspondence to a logit model as in the static case.

A further remark is in order here. It would be interesting to investigate also which kind of micro dynamic economic theory (analogous to a dynamic entropy) may give rise to a dynamic logit model; or alternatively: which is the solution of a general optimal control problem describing a general spatial interaction system of which system (5.7) is a particular case? The latter topic has been discussed in Nijkamp and Reggiani (1988b) who showed how particular forms of urban dynamics, (i.e., decline), may emerge as an equilibrium solution of a general optimal control model of type (5.7) in which the objective function has been replaced by a general well behaved (collective) utility function.

Finally, a remark on the use of optimal control theory in SIMs may be made here. Optimal control theory, and dynamic programming in general, is a common tool in economics and operations research (see, among others, Isard and Liossatos, 1979, Kamien and Schwartz, 1981, Miller, 1979, and Nijkamp, 1980), but relatively seldom has optimal control theory been applied to SIMs (see Slater, 1988, Tan and Bennet, 1984, and Wilson, 1981). Consequently, the present study sheds a new light on spatial interaction models by showing also the potential of this mathematical tool, especially - as will be illustrated in subsequent chapters - in combination with multiobjective analysis, stochasticity and chaos theory.

5.3 Spatial Interaction Models Analyzed by Means of Stochastic Optimal Control

5.3.1 Introduction

In the previous section we have focussed attention on dynamic aspects of SIMs with particular reference to the connection with DCMs. However, analogous to the static context, also in a dynamic framework various attributes of models are not always fully deterministic but subject to stochastic disturbances. Thus it is worth exploring the impact of random disturbances in the behaviour of dynamic SIMs.

In general, introduction of random noise in spatial systems reflects the fact that time series of many spatial economic variables exhibit stochastic fluctuations. There are unpredictable - **exogenous** - events which influence spatial-economic activities at both the macro level (e.g., storms, earthquakes, wars) and the meso level (e.g., strikes, communication delays, weather conditions, economic-political changes), and indirectly at the micro level where also individual idiosyncrasies have to be taken into account in the analysis of factors producing variations around the expected value of flows in the time period considered.

Moreover, such fluctuations, even of small intensity, can sometimes drive the system far away from equilibrium thus causing dramatic changes. Consequently, it is necessary to study, in a dynamic analysis, the possibility of fluctuations as well as their impact on the whole system. In this context it is important to add that fluctuations may also emerge **endogenously** (i.e., without external shocks) in a system whose time dependence is deterministic (i.e., there exists an in-built driving force, either in terms of differential or difference equations, for calculating the behaviour of the system from given initial conditions). This latter analysis will be dealt with in chaos theory, to be shown in Chapter 6 and subsequent chapters, while in the present chapter we will analyze the effects of fluctuations emerging from exogenous shocks on spatial systems, with particular reference to SIMs. For this purpose we will examine here the possibility of dealing with stochastic movements by means of a rather novel approach in spatial analysis - though well-known in economics and finance - viz. the so-called Brownian motion (or Wiener process).

By defining as stochastic processes "systems in which a certain time-dependent random variable X(t) exists" (cf. Zhang, 1991), we note that stochastic processes in spatial analysis have often been associated with Markov chain analysis (in which the knowledge about present situations determines the future) and hence with transition probabilities of one state to another. In this framework a particular type of equation which has been successfully applied in scientific research is the master equation approach (see Gardiner, 1983, Nicolis and Prigogine, 1977, and Weidlich and Haag, 1983) belonging to the broad family of compartmental analysis (see, for a review, De Palma and Lefèvre, 1987, and Leonardi and Campisi, 1981). In general, compartmental analysis describes the evolution of a system by subdividing it into a finite number of subsystems (i.e., the compartments) between which the fundamental units of the system move. Master equations have often been applied in their linear form to ecology (see, e.g., Matis and Hartley, 1971), to interregional mobility and to intraurban movement decisions (see Bartholomew, 1973, and Ginsberg, 1972) and in their non-linear form to migration systems (Haag, 1989a, Haag and Weidlich, 1984, 1986, and Kanaroglou et al., 1986 a, 1986b), to transportation analysis (Kahn et al., 1981, 1982), and to residential location (Ben-Akiva and De Palma, 1986, De Palma and Lefèvre, 1985, and Rabino and Lombardo, 1986). We recall here that some of these studies also link compartmental analysis of a master equation type to discrete choice analysis of a logit type from both a theoretical point of view (see, for example, Barentsen and Nijkamp, 1988, De Palma and Lefèvre, 1983, Haag, 1986, Kanaroglou et al., 1986a, 1986b, and Leonardi, 1985b) and an empirical point of view (see, in particular, the application of a generalized nested MNL model to interprovincial migration in the Netherlands by Liaw and Schuur, 1988).

However, we have to note that - despite their interesting formal rigour - master equations essentially deal with macroscopic variables by using a phenomenological probabilistic description of the microscopic world. Thus the aggregation process may neglect microscopic processes which could give rise to fluctuating forces driving the systems far from equilibrium.

Hence there is a need, as noticed before, to define the dynamics of aggregate variables, by considering such microscopic fluctuating forces as random noises satisfying appropriate requirements. In other words, we may then add a stochastic term, with canonical properties related to standard diffusion processes, to the differential equation describing the evolution of a state variable. Such an approach in a spatial analysis context has been proposed so far only by Campisi (1986) for the analysis of an ecological system of a prey-predator type, by Sikdar and Karmeshu (1982) for a non-linear gravity model applied to population growth and by Vorst (1985) for an urban retail model. In the next section we will proceed somewhat further by introducing random fluctuations in the migration model described in Section 5.2.2, with the purpose of examining their impact in the framework of an optimal control SIM.

5.3.2 A stochastic optimal control approach

Many random phenomena in economic, physical and biological problems can well be approximated by stochastic processes called Gaussian processes, which have very simple analytical distributions. Note that as a stochastic process we define here (following Arnold, 1974, p. xi) "a differential equation for random functions", so that dynamics is implicitly embodied in the concept of stochasticity.

Brownian motions or Wiener processes are a specific kind of a Gaussian process and are widely used for modelling white noise phenomena. It is necessary to underline here that white noise will be considered (following again Arnold, 1974, p. xi) as "a very useful mathematical idealization for describing random influences that fluctuate rapidly and hence are virtually uncorrelated for different instants of time".

In particular for a Wiener process z and for any partition t_0, t_1, t_2 of the time interval, the random variables $z(t_1) - z(t_0)$, $z(t_2) - z(t_1)$, $z(t_3) - z(t_2)$, ... (i.e., the incremental changes) are independently and normally distributed with mean zero and variances $t_1 - t_0$, $t_2 - t_1$, $t_3 - t_2$, ..., respectively (see Kamien and Schwartz, 1981, Kushner, 1971, and Malliaris and Brock, 1982).

If we consider now the evolution of the outflows O_i, previously discussed in equation (5.3), we can incorporate stochasticity of a white noise type in order to take into account stochastic factors in population growth such as birth, death, immigration and emigration rates (cf. also Sikdar and Karmeshu, 1982). This leads to a **stochastic formulation** which would be more realistic since it incorporates **environmental fluctuations**.

Consequently we consider the following stochastic version of the non-linear migration model (5.3), i.e.,:

$$\begin{aligned} dO_i &= \tau_i O_i + \phi_i \left(\sum_j T_{ji} - \sum_j T_{ij}\right) dt + O_i \sigma_i dz_i \\ &= g_i dt + O_i \sigma_i dz_i \end{aligned} \tag{5.12}$$

where τ_i and ϕ_i are growth rates.

The last term at the right hand side of equation (5.12) represents the total stochastic perturbating force which is proportional to the outflow size and can be regarded as "the sum total of various randomly fluctuating environmental factors" (cf. Sikdar and Karmeshu, p. 586). Consequently, it is plausible to argue that this total effect of large numbers of random factors, by the central limit theorem, obeys the Gaussian law. Furthermore, also the time scale of the fluctuations, being much smaller than the macroscopic time scale of outflow change, can be represented by white noise (see again Sikdar and Karmeshu, 1982). Therefore, in (5.12) dz_i is the incremental change (white noise) in a stochastic process z that satisfies a Brownian motion (as previously described).

Such a stochastic formulation in a SIM may be relevant, as movements of people are never taking place in the same context or environment, but are influenced by external perturbations (e.g., weather conditions) or intra-systemic synergies (e.g., congestion).

Obviously in (5.12) each variable is dependent on time; the symbol t has been omitted for the sake of simplicity. Equation (5.12) is here a stochastic differential equation which (with initial condition $O_i(0) = O_i^*$) can be solved by means of new rules of differentiation and integration, such as the ones provided in Itô's stochastic calculus (cf. Arnold, 1974). In particular, by using Itô's theorem, it can be shown (see Nijkamp and Reggiani, 1989b) that the solution of (5.12) is the following:

$$O_i = F(t,z) = O_i^* \exp(-\phi_i - \sigma_i^2/2)t + \sigma_i z_i + [\tau_i + (\phi_i / O_i) \sum_j T_{ji}]t \tag{5.13}$$

In (5.12) g_i represents the expected rate of change (i.e., the drift), while σ_i is the diffusion component of the stochastic process; clearly, in a deterministic context, $\sigma_i = 0$. It should be noted here that besides Itô's rule other approaches can be used for solving stochastic differential equations, such as the Stratonovich calculus (cf. Arnold, 1974, and Sikdar and Karmeshu, 1982, for a related application to population systems). The global asymptotic stability of these stochastic differential equations can be analyzed by means of the Liapunov-Kushner approach (cf. Malliaris and Brock, 1982, and Vorst, 1985, for a related application to retail models).

Here we aim to introduce the **stochastic differential equation** (5.12) in an optimal control SIM; this means the development of a hitherto unexplored approach in spatial interaction analysis. A further step is the introduction of multiple objective functions in order to consider different driving mechanisms of a dynamic spatial interaction system (see Nijkamp and Rietveld, 1986, and Rietveld, 1980); this leads to multicriteria optimal control models. In particular we will use the objective functions already discussed in Section 2.3, so that we will link, at the end, three issues in SIMs, viz., **multiple objectives, spatial dynamics and stochasticity.**

Consequently we assume two objective functions (viz., an entropy function and a cost function as described in (2.20)), so that we may finally design the following optimal control model, in which the mathematical expectation E of the weighted objective functions has to be maximized:

$$w = \max E \int_0^{\bar{T}} - [\alpha_i \sum_i \sum_j T_{ij} (\ln T_{ij} - 1) + \alpha_2 \sum_i \sum_j c_{ij} T_{ij}]dt$$

subject to

$$\sum_j T_{ij} = O_i$$

$$\sum_i T_{ij} = D_j \tag{5.14}$$

$$dO_i = \tau_i O_i + \phi_i (\sum_j T_{ji} - \sum_j T_{ij})dt + O_i \sigma_i dz_i$$

$$= g_i dt + O_i \sigma_i dz_{i2}$$

It can be shown that the solution of the stochastic optimal control problem (5.14) is the following (see Annex 5B):

$$T^*_{ij} = A^*_i O_i Z_i B^*_j D_j \exp(-\beta c_{ij}) \tag{5.15}$$

where:

$$A^*_i = A_i \exp(-\phi_i \Phi^*_i / \alpha_1) \tag{5.16}$$

$$B^*_j = B_j \exp(+ \phi_j \Phi^*_j / \alpha_1) \tag{5.17}$$

and:

$$Z_i = \exp(O_i \sigma^2_i \frac{\partial \Phi^*_i}{\partial O_i} / \alpha_1) \tag{5.18}$$

In equations (5.16) and (5.17) A_i and B_j have similar definitions as in (2.23) and (2.24), while Φ_i^* is the stochastic costate variable (see again Annex 5B). Also β has the same meaning as in (2.22). The interesting feature here is that A_i^* and B_j^* are stochastic terms, as they incorporate the costate variable Φ_i^* satisfying (5B.3), (5B.6) and (5B.7) (see Annex 5B). Also the term Z_i is stochastic because it incorporates either Φ_i^* or the diffusion component σ_i^2 of the stochastic Wiener process Z_i. However, owing to the difficulties involved in the solution of (5B.6), an explicit solution of A_i^*, B_j^*, Z_i and consequently of T_{ij}^* is impossible. It can be easily seen that the stochastic solution (5.15) can formally be transformed into a logit expression as follows:

$$P_{ij}^* = \frac{T_{ij}^*}{O_i} = Z_i \frac{B_j^* D_j \exp(-\beta c_{ij})}{\sum_j B_j^* D_j \exp(-\beta c_{ij})} \tag{5.19}$$

The analytical (stochastic) form (5.19) is in agreement with the standard results from conventional SIMs (see Section 5.2); in fact when $\sigma_i = 0$, equation (5.19) yields as a special case the MNL model (5.11), obtained from the deterministic approach. Consequently, introduction of stochastic processes of a white noise type in SIMs offers a generalized MNL model solution including stochastic fluctuations; in particular, it is noteworthy that stochastic fluctuations tend to destabilize a SIM owing to the exponential form of σ_i^2 implied by Z_i. This result bears some resemblance to external fluctuations in biological models, where the impact of the environmental fluctuations implies a change in the density of a population in an ecosystem with reference to the level valid in a deterministic context (see, for example, Maynard Smith, 1974). In particular, in case of strong fluctuations, the population can become extinct. Besides, there is a close link to the result emerging from the stochastic analysis of a SIM, proposed by Sikdar and Karmeshu (1982); in fact also these authors show that the stochastic perturbations tend to destabilize their model.

Moreover we can also draw a parallel between the final solution (5.19) and the transition probability considered by Cordey-Hayes and Gleave (1974) (see also Pickles, 1980); in fact we can decompose expression (5.19) into an 'escape frequency' (depending on the variance σ_i^2, which may be caused by age distribution of population, variation of income with time, etc.) and into a 'capture probability' (determined by the stochastic logit form) which models the probability of moving toward j after escaping from i. Obviously in this context i and j represent an origin and destination state. Some similarity can also be found in the kinetic theory of urban migration, developed by Tomlin (1979).

5.3.3 Concluding remarks

In this section we have shown how the solution of a stochastic optimal control SIM can offer a new interpretation in choice processes, i.e., a two-level structure which is the result of two factors: the stochastic possibility of moving (e.g., migrating) and the (stochastic) destination choice probability (see again Liaw and Schuur, 1988). This is in synthesis expressed by the analytical form (5.19).

Another interesting remark emerging from the present analysis is that solution (5.19) is formally compatible with a logit expression; this confirms also the descriptive and explanatory power of logit models and consequently of macro approaches related to the maximization of entropy in a dynamic context (see also the previous section). It should also be noted that the use of multi-objective analysis in a stochastic context is interesting; multi-objective analysis has been used several times in dynamics (see for instance Nijkamp, 1977, 1979b, and 1980), but it has rarely been linked to stochasticity (see, for example, Ermoliev and Leonardi, 1981).

In the next section the same kind of approach will be used for analyzing the possibility of catastrophe behaviour in dynamic SIMs.

5.4 Spatial Interaction Models with Catastrophe Behaviour Analyzed in the Framework of Stochastic Optimal Control

5.4.1 The model

In this section we will pay attention to a production-constrained SIM by extending this analysis in a stochastic framework. In particular we will start with the production-constrained SIM originally developed by Huff (1964) and Lakshmanan and Hansen (1965) in the context of the urban retail sector. Here, for the sake of uniformity with respect to the analysis presented in the previous section, we will consider a migration context. We assume the following model:

$$T_{ij} = O_i \frac{W_j^a \exp(-\beta c_{ij})}{\sum_{j} W_{j'}^a \exp(-\beta c_{ij'})} \qquad j,j' = 1,\dots,J \tag{5.20}$$

where:

T_{ij} is the volume of people immigrating from their zone of origin i to zone j;

W_j is the number of workplaces in zone j;

O_i is the total volume of outflows from origin i;

c_{ij} is the migration distance between living zone i and labour market j;

a is a parameter reflecting the degree of immigrants' scale economies ($a > 0$);

β is the usual deterrence parameter.

The production-constrained SIM, which in the formulation (5.20) shows the link with a logit structure, has already been discussed in Chapter 1 and Chapter 4.

Following Harris and Wilson's analysis (1978) related to formulation (5.20) for shopping activities, we can add the hypothesis that the number of workplaces W_j will develop until it is in balance with the total inflows $\sum_i T_{ij}$, i.e.,:

$$\sum_i T_{ij} = K W_j \qquad j = 1,\dots,J \tag{5.21}$$

where K can be interpreted as an accessibility parameter converting inflow units into workplace units.

The equilibrium condition (5.21) may be assumed to arise from the differential equation (5.1), i.e.,:

$$\frac{dW_j}{dt} = \dot{W}_j = \nu \left(\sum_i T_{ij} - K W_j\right) \tag{5.22}$$

where ν is the response rate of the system.

Equation (5.22) assumes that W_j may grow if $\sum_i T_{ij} > K W_j$ and decline if $\sum_i T_{ij} < K W_j$. This may be plausible for example in the case of growing cities. The equilibrium conditions (5.20) and (5.21) have been thoroughly analyzed by Harris and Wilson (1978) and Wilson (1981); it is easy to see that for $a \geq 1$ a catastrophe behaviour may emerge. This may be explained by the fact that an increase in immigration leads to a rapid increase in the workplaces, thus causing possible unstable behaviour (e.g., due to congestion phenomena).

In the next section we will show that the equilibrium conditions (5.20) and (5.21) may also emerge, in their analytical formulation, from a stochastic optimal control SIM. Consequently also the possibility of a catastrophe behaviour in a stochastic context will be discussed.

5.4.2 The stochastic optimal control version

As we did in the previous section, we want to analyze the production-constrained SIM by adding a stochastic term to the differential equation (5.22) which describes its dynamics. This stochastic term will take into account the stochastic factors related to the growth of the workplaces. Thus we have the following stochastic differential equation:

$$\begin{aligned} dW_j &= [\nu (\sum_i T_{ij} - K W_j)] \, dt + W_j \, \sigma_j \, dz_j \\ &= \bar{g}_i \, dt + W_j \, \sigma_j \, dz_j \end{aligned} \tag{5.23}$$

where $\bar{g}_i$ represents the spatial 'drift' and where the last term is the stochastic perturbating force (proportional to the number of workplaces). Thus σ_i is the diffusion component and dz_j is the incremental change (white noise) in a stochastic process z_j satisfying a Brownian process, as we discussed in Section 5.3.

As mentioned before, a stochastic differential equation similar to (5.23) has already been analyzed by Vorst (1985) in a shopping centre context. We recall that in the absence of a stochastic component, assumption (5.23) implies an upward pressure on the number of workplaces, if the attractiveness of j increases (that is, if the capacity $\sum_i T_{ij}$ is growing) and a downward pressure if the labour force is growing (that is, if W_j is increasing). As a second step we will introduce the stochastic differential equation (5.23) in an optimal control framework. For this purpose we will assume as an objective function a generalization of Wilson's (1981) specification, viz., a bi-objective function of the following type:

$$u = \max E \int_0^{\bar{T}} [-\frac{1}{\beta} \sum_i \sum_j T_{ij} (\log T_{ij} - 1) + \sum_i \sum_j T_{ij} (\frac{a}{\beta} \log W_j - c_{ij})] \, dt \tag{5.24}$$

Formulation (5.24) maximizes essentially the mathematical expectation E of the weighted objective function; the first summation term in square brackets represents the entropy (interaction) of our spatial system, weighted by means of the coefficient $1/\beta$, whereas the second term denotes the aggregate net benefits of people living in i and immigrating to j (with a as a scale parameter). Altogether the objective function resulting from (5.24) is again a multiperiod cumulative utility function (see the previous sections). Therefore we have finally the following optimal control model (with T_{ij} as a control variable):

$$u = \max E \int_0^{\bar{T}} [-\frac{1}{\beta} \sum_i \sum_j T_{ij} (\log T_{ij} - 1) + \sum_i \sum_j T_{ij} (\frac{a}{\beta} \log W_j - c_{ij})] dt$$

subject to (5.25)

$$\sum_j T_{ij} = O_i$$

$$d W_j = \bar{g}_j \, dt + W_j \, \sigma_j \, dz_j$$

The solution of problem (5.25), which follows directly from the solution of the slightly different problem (5.14) (derived in Annex 5A), is straightforward. In particular the optimal solution T^*_{ij} (see also Nijkamp and Reggiani, 1988c) is the following:

$$T^*_{ij} = A^*_i \, O_i \, W^a_j \exp(\beta \nu \Phi^*_j - \beta c_{ij}) \tag{5.26}$$

where

$$A_i = 1 / \sum_j W^a_j \exp(\beta \nu \Phi^*_j - \beta c_{ij}) \tag{5.27}$$

and Φ^*_j is the stochastic costate variable Thus, owing to the term Φ^*_j, T^*_{ij} is a stochastic expression, incorporating a social cost-benefit measure; the term $\beta v \Phi^*$ reflects the (stochastic) shadow price of employment growth in j, whilst the term βc_{ij} reflects the distance friction costs. The parameter β is related to the perception of immigrants for bridging the distance between zone i and zone j.

Then, it is interesting to observe that if we impose the boundary condition $\Phi_j^*(\tilde{T}) = 0$, we find for $t = \tilde{T}$ the original Wilson model described in (5.20):

$$T_{ij} = O_i \frac{W_j^a \exp(-\beta c_{ij})}{\sum_j W_j^a \exp(\beta c_{ij})} \tag{5.28}$$

In this case (for $t = \tilde{T}$), when we also suppose that the deterministic part of equation (5.23) is in equilibrium, we get:

$$\sum_i T_{ij} = K W_j \tag{5.29}$$

and consequently:

$$dW_j = W_j \sigma_j dz_j \tag{5.30}$$

It is easy to see (see Reggiani, 1990) that the solution of (5.30) is:

$$W_j = W_j^o \exp\left(-\frac{\sigma_j^2}{2} t + \sigma_j z_j\right) \tag{5.31}$$

which represents the optimal trajectory of the state variable W_j under the equilibrium condition (5.29).

It should be noted that (5.28) and (5.29) are exactly the equilibrium conditions (5.20) and (5.21) analyzed in the deterministic context; the only difference here is that W_j is stochastic, as we can see from (5.31).

Therefore, if we represent equations (5.28) and (5.29) in the plane $(\sum_i T_{ij}, W_j)$, we have clearly the possibility of both stable and unstable points, while in the plane (W_j, K) we have the possibility of catastrophe behaviour (see Annex 5C).

In particular, in this latter case, we can see that there is a critical value (K^c) beyond which the number of workplaces jumps to zero (see Figure 5C.2). Since K is an accessibility parameter converting inflows into workplaces, we can - also under our hypothesis of a stochastic W_j - interpret this catastrophic change on the basis of the phenomenon of congestion, like in the deterministic context. In fact, from (5.31) we can see that the diffusion components σ_j do not influence this dramatic change, since the only way to obtain the condition $W_j = 0$ is a sudden decrease to zero of the initial conditions W_j^o.

The conclusion is that the inclusion of stochastic perturbances in the singly-constrained dynamic SIM affects the optimal trajectories, so that we get again stochastic movements stemming from deterministic equations. However, if we have conditions for the appearance of catastrophe behaviour, the stochastic disturbances do not necessarily drive these drastic changes toward zero values of the state variable. This interesting result hence shows the characteristic of **structural stability** of dynamic SIMs emerging from an optimal control formulation, meaning that their qualitative properties are not lost when they are subject to small perturbations.

5.5 Epilogue

This chapter has focussed attention on dynamic and stochastic aspects of SIMs with particular reference to the connections with DCMs.

In particular, we emphasize the important finding that also in a dynamic context an analytical correspondence exists between SIMs and MNL models. This result has been obtained in the first part of the present chapter by applying an optimal control approach to a dynamic entropy model.

A first remark related from the previous result is the important role of an entropy relationship in a dynamic context; the relevance of entropy has also been shown in the second part of this chapter where its use has again led to clear analytical solutions that are compatible with SIMs. This opens the important issue regarding the connections between entropy and SIMs as well as their capability in examining order and disorder of real-world phenomena.

A further remark concerns the power of mathematical tools (such as optimal control theory) in a spatial interaction context in order to provide new insights into complex dynamics of spatial systems.

Then in the second part of this chapter it has been shown that the family of SIMs can also emerge from a stochastic spatial analysis. Firstly, we have studied a stochastic optimal control entropy model in a way parallel to the deterministic problem previously analyzed; the solution appears to be a stochastic movement compatible with a doubly-constrained SIM.

Secondly, we have concentrated our attention on the stochastic extension of a singly-constrained SIM of the Wilson type; even though the solution is again a stochastic movement stemming from a deterministic equation, under the conditions of a catastrophe behaviour the stochastic disturbances do not affect the evolution of the dramatic changes.

The main conclusion is that SIMs have the possibility of representing a wide spectrum of interaction activities (and consequently of controlling spatial systems) by means of a new stochastic (i.e., subject to random phenomena) formulation.

In the framework of these observations a main problem remains to be studied, viz. the possibility of empirically verifying the above mentioned result that the equilibrium point governing an optimal entropy system is a dynamic (or stochastic) SIM (or MNL model). Owing to the lack of suitable empirical time series data, this aspect still remains to be investigated, so that for the time being we have to take resort to simulation experiments.

A final general remark emerging from the stochastic movements resulting from our analysis and concealed in the deterministic framework is the more important question whether there exist types of models capable of generating complexity in dynamic phenomena while retaining extreme simplicity in their structure. It can be shown that chaotic models in general, and ecologically-based models in particular, related to SIM analysis, provide interesting possibilities; they will be analyzed in the next chapters.

Annex 5A

The Generalized Spatial Interaction Model as a Solution to the Optimal Control Entropy Model

In this annex we will investigate the solution of the following optimal control entropy model, already discussed in Section 5.2, i.e.:

$$\max w = \int_0^{\bar{T}} - \sum_i \sum_j T_{ij} (\ln T_{ij} - 1)\, dt$$

subject to

$$\begin{aligned} & \sum_j T_{ij} = O_i \\ & \sum_i T_{ij} = D_j \\ & \sum_i \sum_j T_{ij}\, c_{ij} = C \\ & \dot{O}_i = \tau_i O_i + \phi_i \left(\sum_j T_{ji} - \sum_j T_{ij}\right) \end{aligned} \tag{5A.1}$$

Owing to the presence of static constraints on O_i, D_j and C, problem (5A.1) is a bounded optimal control model (see e.g., Miller, 1979, Kamien and Schwartz, 1981, and Tan and Bennet, 1984), so that we have to use both the Hamiltonian and the Lagrangian function for the derivation of the optimal solution. The Hamiltonian H is:

$$H(T_{ij}, O_i, t) = - \sum_i \sum_j T_{ij} (\ln T_{ij} - 1) + \sum_i \Phi_i \dot{O}_i \tag{5A.2}$$

with Φ_i the costate variable, related to $\dot{O}_i$, which can be interpreted as a dynamic accessibility factor to zones i (see Section 5.2). The Lagrangian L is:

$$\begin{aligned} L(T_{ij}, O_i t) = {} & H(T_{ij}, O_i, t) + \sum_i \lambda_i (O_i - \sum_j T_{ij}) + \\ & + \sum_j \mu_j (D_j - \sum_i T_{ij}) + \beta (C - \sum_i \sum_j c_{ij} T_{ij}) \end{aligned} \tag{5A.3}$$

where λ_i, μ_j and β are the Lagrange multipliers associated with the static constraints in (5A.1).

The necessary first-order conditions for optimality are:

$$\frac{\partial L}{\partial T_{ij}} = 0$$

$$\frac{\partial L}{\partial O_i} = \dot{\Phi}_i \qquad (5A.4)$$

$$\frac{\partial L}{\partial \Phi_i} = O_i$$

Since the integrand in (5A.1) is concave, i.e.:

$$\frac{\partial}{\partial T_{ij}} \left(\frac{\partial L}{\partial T_{ij}}\right) = \frac{-1}{T_{ij}} < 0 \qquad (5A.5)$$

The conditions (5A.4) are also sufficient for optimality. Therefore from the first condition of (5A.4) we get:

$$\frac{\partial L}{\partial T_{ij}} = -\ln T_{ij} - \lambda_i - \mu_j - \beta c_{ij} - \phi_i \Phi_i + \phi_j \Phi_j = 0 \qquad (5A.6)$$

and from (5A.6) we can derive T_{ij}:

$$T_{ij} = \exp(-\lambda_i - \phi_i \Phi_i) \exp(-\mu_j + \phi_j \Phi_j) \exp(-\beta c_{ij}) \qquad (5A.7)$$

We can now substitute equation (5A.7) into the constraints of our problem (5A.1):

$$\sum_j T_{ij} = O_i \qquad (5A.8)$$

$$\sum_i T_{ij} = D_j \qquad (5A.9)$$

so that we obtain:

$$\exp(-\lambda_i - \phi_i \Phi_i) = O_i \left[\sum_j \exp(-\mu_j + \phi_j \Phi_j - \beta c_{ij})\right]^{-1} \qquad (5A.10)$$

and:

$$\exp(-\mu_j + \phi_j \Phi_j) = D_j [\sum_i \exp(-\lambda_i - \phi_i \Phi_i - \beta c_{ij})]^{-1} \tag{5A.11}$$

By defining now:

$$A_i^* = \exp(-\lambda_i - \phi_i \Phi_i) / O_i \tag{5A.12}$$

$$B_j^* = \exp(-\mu_j + \phi_j \Phi_j) / D_j \tag{5A.13}$$

the optimal value (5A.7) becomes:

$$T_{ij} = A_i^* B_j^* O_i D_j \exp(-\beta c_{ij}) \tag{5A.14}$$

It is clear that the dynamic balancing factors A_i^* and B_j^* have the following expressions, obtained from equations (5A.8 - 5A.13):

$$A_i^* = \{\sum_j B_j^* D_j \exp(-\beta c_{ij})]^{-1} \tag{5A.15}$$

$$B_j^* = \{\sum_i A_i^* O_i \exp(-\beta c_{ij})\}^{-1} \tag{5A.16}$$

Consequently, the final result (5A.14) represents a **generalized SIM**.

We now have to analyze the conditions for the state and co-state variables, since the optimal control solution for T_{ij} still contains the (unknown) state variables O_i. The optimality conditions for the state variables are on the basis of (5A.4):

$$\frac{\partial L}{\partial O_i} = \dot{\Phi}_i \tag{5A.17}$$

$$\frac{\partial}{\partial O_i} \{-\sum_i \sum_j T_{ij} (\ln T_{ij} - 1)\} + \Phi_i (\tau_i - \phi_i) + \lambda_i = -\dot{\Phi}_i \tag{5A.18}$$

which together with the previous conditions can in principle be solved numerically.

Therefore, it is evident from this annex that the optimal path related to an optimal control entropy model is a (dynamic) SIM (in the generalized formulation (5A.14)) or a (dynamic) MNL model. Moreover, it is interesting to notice that the interrelations between entropy and SIMs (as observed in Chapter 2) also exist in a dynamic framework.

Annex 5B

A (Generalized) Stochastic Spatial Interaction Model as a Solution to a Stochastic Optimal Control Model

In this annex we will illustrate the solution stemming from our stochastic optimal control model (5.14). To this end we will analyze model (5.14) by first regarding the constraints on the control variables T_{ij}. Following Malliaris and Brock (1982), we may write the Hamilton-Jacobi-Bellman equation associated with (5.14) as:

$$-\frac{\partial w}{\partial t} = \max_{T_{ij}} \Big(- [\alpha_1 \{\sum_i \sum_j T_{ij} (\ln T_{ij} - 1)\} + \alpha_2 \sum_i \sum_j c_{ij} T_{ij}] + \sum_i \frac{\partial w}{\partial O_i} g_i + 1/2 \sum_i \frac{\partial^2 w}{\partial O_i^2} (\sum_j T_{ij})^2 \sigma_i^2 \tag{5B.1}$$

where the assumption is made that O_i is uncorrelated with O_j. Equation (5B.1) can now be written as:

$$\frac{\partial w}{\partial t} = \max_{T_{ij}} H* \tag{5B.2}$$

where H* is the functional form of the expression in brackets at the right hand side of (5B.1).

Next, if we define the costate variable $\Phi_i^*(t)$ as follows:

$$\Phi_i^*(t) = \frac{\partial w}{\partial O_i} \tag{5B.3}$$

we may write H^* as:

$$H* = - [\alpha_1 \{\sum_i \sum_j T_{ij} (\ln T_{ij} - 1)\} + \alpha_2 \sum_i \sum_j c_{ij} T_{ij}] + \sum_i \Phi_i^* g_i + 1/2 \sum_j \frac{\partial \Phi_i^*}{\partial O_i} \sigma_i^2 (\sum_j T_{ij})^2 \tag{5B.4}$$

Now we will introduce constraints on the control variables. Then it is straightforward (see Chow, 1979) to define the following Lagrangian expression:

$$L^* = H^* + \sum_i \lambda_i (O_i - \sum_j T_{ij}) + \sum_j \mu_j (D_j - \sum_j T_{ij}) + m(\sum_j D_j - \sum_i O_i) \qquad (5B.5)$$

where λ_i, μ_j and m are the Lagrange multipliers related to the static constraints.

In the latter case we may apply the Pontryagin Stochastic Maximum Principle (see Malliaris and Brock, 1982). This principle states that for an optimal control variable T^*_{ij} that maximizes the Lagrangian (5B.5) the following conditions hold:

- the costate function Φ^*_i satisfies the following stochastic differential equation:

$$d\Phi^*_i = - \frac{\partial L^*}{\partial O_i} dt + \sum_i \frac{\partial \Phi^*_i}{\partial O_i} O_i \sigma_i dz_i \qquad (5B.6)$$

- the following transversality condition holds:

$$\Phi^*_i \{O_i (\tilde{T}), \tilde{T}\} = \frac{\partial w}{\partial O_i} \{O_i (\tilde{T}), \tilde{T}\} \geq 0$$

$$\Phi^*_i (\tilde{T}) O_i (\tilde{T}) = 0 \qquad (5B.7)$$

Furthermore, it is easily seen that the optimal solution T^*_{ij} is equal to:

$$T^*_{ij} = \exp \{(- \lambda_i - \mu_j - \phi_i \Phi^*_i + \phi_j \Phi^*_j - \alpha_2 c_{ij} + O_i \sigma^2_i \frac{\partial \Phi^*_i}{\partial O_i}) / \alpha_1\} \qquad (5B.8)$$

This expression leads thus to the interesting result that - apart from the last stochastic term - the same formal outcome is obtained as in Annex 5A, so that the final solution is:

$$A^*_i = A_i \exp(- \phi_i \Phi^*_i / \alpha_1) \qquad (5B.9)$$

with:

$$T^*_{ij} = A^*_i O_i Z_i B^*_j D_j \exp(- \beta c_{ij}) \qquad (5B.10)$$

$$B_j^* = B_j \exp(\phi_j \Phi_j^* / \alpha_1) \tag{5B.11}$$

$$Z_i = \exp (O_i^2 \sigma_i \frac{\partial \Phi_i^*}{\partial O_i} / \alpha_1) \tag{5B.12}$$

and

$$\beta = \alpha_2/\alpha_1 \tag{5B.13}$$

$$A_i = \exp(- \lambda_i/\alpha_1) / O_i \tag{5B.14}$$

$$B_i = \exp(- \mu_j / \alpha_1) / D_j \tag{5B.15}$$

Consequently, in a deterministic context where $\sigma_i = 0$ we find that solution (5B.8) coincides exactly with the usual dynamic SIM expressed in equation (5A.14).

Annex 5C

Stability and Bifurcations in a Phase Diagram Analysis for a Stochastic Optimal Control Problem

The equilibrium solutions (at $t = \tilde{T}$) (5.28) and (5.29) can be easily written, by following Harris and Wilson (1978):

$$\sum_i T_{ij} = K\, W_j \tag{5C.1}$$

$$\sum_i T_{ij} = \sum_i \frac{O_i\, W_j^a \exp(-\beta c_{ij})}{\sum_j W_j^a \exp(-\beta c_{ij})} \tag{5C.2}$$

where the state variable W_j is a stochastic function obeying equation (6C.3) and T_{ij} is the control variable. However, if we consider a phase diagram analysis (by eliminating t and z), we can easily deduce the functional form of the inflow $\sum_i T_{ij}$ towards the workplaces W_j for different values of a (see Figure 5C.1).

If in a way analogous to the urban retail model of Harris and Wilson, we consider the effect of varying K, we obtain a set of equilibrium points linked to the fold catastrophe (see Figure 5C.2).

It is clear that there are values of $K(K^c)$, beyond which W_j will jump from W_j^A to zero (for $a \geq 1$). This means that for the case $a \geq 1$ there is a critical value K^c in the accessibility beyond which the number of workplaces jumps to zero. This fact can be explained, for example, through the phenomenon of congestion. Therefore it is only for $a < 1$ that we do not have a catastrophe behaviour, because in this case there is a unique positive equilibrium (see also Kaashoek and Vorst, 1984). It should be noted that also changes in β might lead to possible jumps in the number of workplaces (see Wilson, 1981).

The conclusion is that at time $t = \tilde{T}$, and under the condition that the deterministic part of equation (5.23) is in equilibrium, certain smooth parameter changes may lead to discrete changes in the state variable W_j.

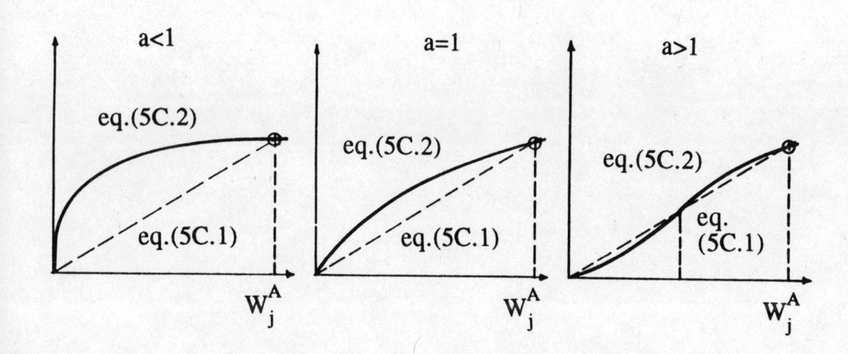

Figure 5C.1 The Inflow Curves and the Equilibrium Points (W_j^A represents stable points and W_j^B represents unstable points)
y-axis = $\sum_i T_{ij}$; x-axis = W_j

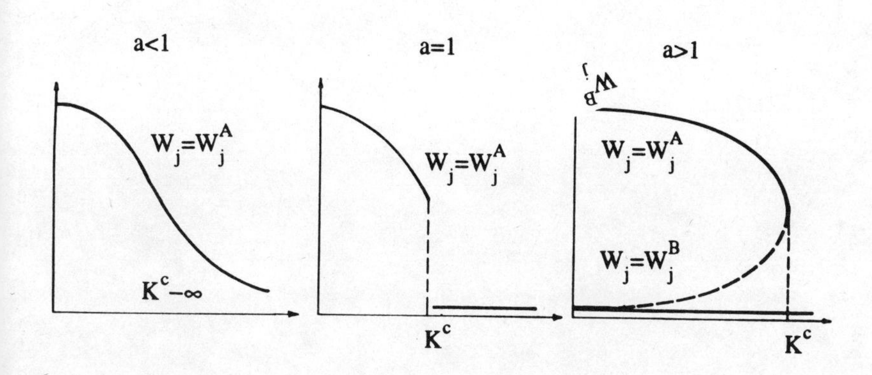

Figure 5C.2 The Equilibrium Points as a Function of K
y-axis = W_j; x-axis = K

CHAPTER 6
SPATIAL MODELLING AND CHAOS THEORY

6.1 Prologue

In the previous chapter we have shown that a SIM which takes into account external stochastic influences is highly dependent on the diffusion component (i.e. the amplitude) of these exogenous forces. However, as mentioned before, various types of fluctuations may influence spatial-economic behaviour, and some of them may be 'governed' by deterministic impact mechanisms. In other words, such irregular oscillations may be created **endogenously** through a non-linear lag structure in the behaviour of a system and described by relatively simple differential equations. This second way of explaining - seemingly stochastic - endogenous fluctuations will be treated in the present chapter by means of concepts stemming mainly from **chaos theory**.

Given the relevance of chaotic phenomena in economics and other social sciences, in the first part of this chapter much attention will be devoted to some important foundations of chaos as well as to some applications of chaos theory in the specific field of spatial-economic analysis. The second part of the chapter will then give an illustration of the role of chaos in dynamic spatial systems by investigating the relationships between chaotic movements and urban dynamics. In particular a generalized Lorenz model will be adopted here for explaining urban decline. Simulation experiments will also be carried out in order to analyze the complex behaviour emerging from this model in relation to different values of the parameters and of the initial conditions. The main conclusion from this experiment seems that in spatial models chaos is a possibility, not so easy to reach. More attention should thus be paid not only to the theoretical structure of chaos models, but also to empirical research testing the range of the parameter values as well as the initial conditions which could lead to chaotic patterns.

6.2 Chaos Theory: A Brief Review

6.2.1 A general introduction to non-linear modelling

Although linear systems are still a prominent research area nowadays, the need for non-linear analysis emerged already in the past century starting from the study of fluid motions in hydrodynamics and aeronautics (see, e.g., Crilly, 1991, for a brief historical background sketch of (non) linear systems).

Linear (or linearized) dynamic systems have elegant mathematical properties. Firstly, outputs are proportional to inputs, so that the complex behaviour of a system can be deduced from knowledge of its constituent parts (the so-called additive property). Secondly, the predictive ability of linear systems demonstrates itself conveniently in empirical studies, since currently popular methods, such as linear regression analysis, can be used.

However, it is clear that in a linear analysis some important and interesting dynamic properties, such as sudden and unexpected discontinuities, multiplicity of (dis)equilibria, bifurcations and catastrophe behaviour, characterizing the dynamics of many real-world phenomena, are missed out. Examples of such types of complex dynamics are inter alia found in the evolution of cities (see, e.g., Mees, 1975), in regional development processes (Andersson and Batten, 1988, and Wilson, 1981), in dynamic choices in spatial systems (Fisher et al., 1990), and of course in a much wider area of general economic analysis (see Turner, 1980). Especially if various subsystems within an economic system (for instance, industry, infrastructure, etc.) are intertwined in a non-linear dynamic way, unexpected switches in evolutionary patterns may take place (see also Nijkamp, 1990).

One of the first interesting examples of catastrophe behaviour applied to spatial systems (and to dynamic SIMs in particular) can be found in the pioneering work by Wilson (1981). More specifically his model (also presented in a stochastic framework in Section 5.4.) shows, in its dynamic logistic form applied to a retail problem, the possibility of multiple equilibria and jumps in the state variable, by slightly varying the parameters (i.e., the control variables) (see also Annex 5C).

Consequently, in recent years there has been an upswing in the interest in **non-linear dynamics**, caused by the increasing awareness of the dynamic interconnecting nature and complex behaviour of real-world phenomena. Thus attention has increasingly shifted from the equilibrium concept towards the notion of 'disequilibrium', 'instability' and 'discontinuities' with a particular view on the driving forces (or conditions) governing transitions (or bifurcations) in a complex dynamic system.

In this context it is necessary to identify the properties and conditions of **structural (in)stability** of a system (see Thom, 1975). As mentioned at the end of Chapter 5, a system is structurally unstable when even small perturbations in functional forms change the qualitative properties of the system. In particular these significant qualitative changes in the behaviour of the system (i.e., in the state variables) are closely connected with bifurcation and catastrophe phenomena which (can) occur as the parameters values (i.e., the control variables) are changed (see for a brief review on bifurcation theory and catastrophe theory applied to spatial-economic patterns, e.g., Dendrinos and Mullally, 1985, and Zhang, 1991). For the sake of simplicity we have provided in Annex 6A a classification scheme of the equilibrium points (stable and unstable) for a non-linear system of two differential equations.

It is clear from the literature that a simple bifurcation can transform the equilibrium points from one state (e.g., stable) to another one (e.g., unstable) (or vice versa). An example in this context is provided in Chapter 8, where we will show the structural instability of a prey-predator system, since a bifurcation (depending on the parameter values) can transform the equilibrium point from a stable 'focus' to an unstable 'focus' via a 'centre'. (The terminology of equilibrium points is illustrated in Annex 6A).

It should be noted that a further analysis, in the study of non-linear dynamics, is offered by the time-dependent behaviour of the system (compared to the phase portrait or the bifurcation diagrams plotting the state variable values against the parameter values). In this context the identification of (a)periodic motions (or oscillations in the state variables) results as a quite interesting outcome.

Periodic or cyclic solutions are very common in economics (see, for example, the wealth of business cycle literature reviewed by Lorenz, 1989, Kelsey, 1988, Puu, 1989, and Zhang, 1991) as well as in ecology (see, e.g., Haken, 1983a, and 1983b), and recently also in geography and regional science (see, e.g., Dendrinos and Mullally, 1985, Nijkamp, 1987a, and Nijkamp and Reggiani, 1990a).

From a mathematical point of view periodic solutions are strictly connected with the existence of **closed orbits** and **limit cycles**. In the first case states are repeated from one orbit to the next; in the second one this is not the case but the orbits are asymptotically close to a closed orbit (see Figure 6.2). In other words, a limit cycle is a closed orbit (emerging as an equilibrium solution from two-dimensional differential equations, see Annex 6A) which is also an attractor (i.e., a bounded region towards which every trajectory of a dynamic system tends to move).

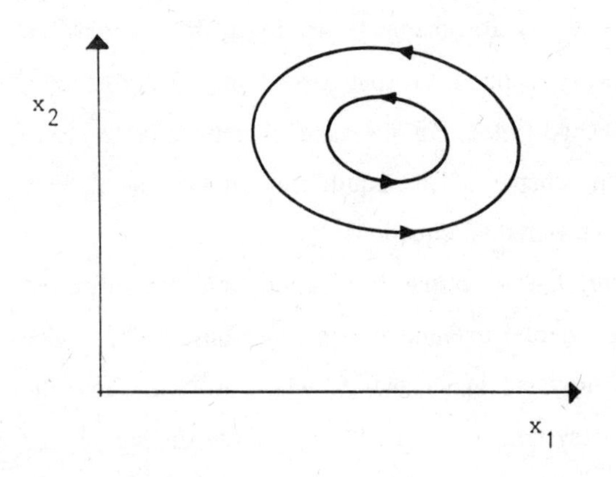

Figure 6.1 Periodic Solutions in Phase Space; a) Closed Orbits

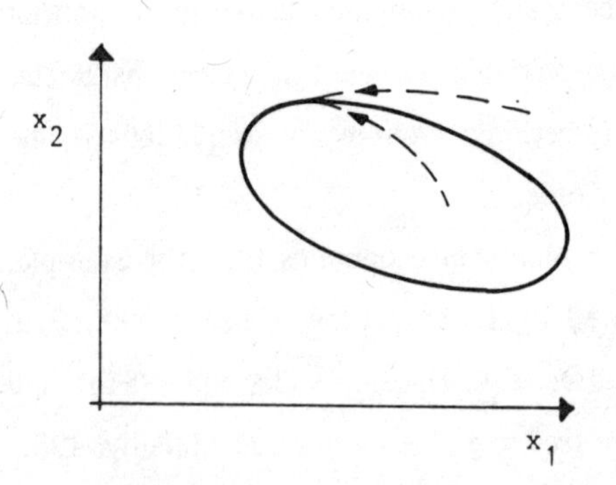

Figure 6.2 Periodic Solutions in Phase Space; b) Limit Cycles

Thus a limit cycle is "closed so that a point moving along the curve will return to its starting position at fixed time intervals and thus execute periodic motion" (cf. Zhang, 1990, p. 207).

It should be noted that there is also the possibility of so-called aperiodic or chaotic oscillations where the amplitude and period vary without any trace of a recognizable pattern. These chaotic motions will be extensively discussed in the next subsection.

A method to find limit cycles in the two-dimensional plane is provided by the Poincaré-Bendixson theorem. This is clarified by Haken (1983a, p. 119) as follows: "Let us pick a point $x^{o} = (x_1^{o}, x_2^{o})$ which is assumed to be non singular and take it as initial value for the solution of the system:

$$\begin{aligned} \dot{x}_1 &= f(x_1, x_2) \\ \dot{x}_2 &= g(x_1, x_2) \end{aligned} \tag{6.1}$$

For later times, $t > t_o$, $x(t)$ will move along that part of the trajectory which starts at $x(t_o) = x^{o}$. We call this part half-trajectory. The Poincaré-Bendixson theorem then states: if a half-trajectory remains in a finite domain without approaching singular points, then this half-trajectory is either a limit cycle or approaches such a cycle. (There are also other forms of this theorem available in the literature)". A consequence of the Poincaré-Bendixson theorem is that the possibility that more than one limit cycle exists is not excluded. Furthermore, if there is more than one limit cycle, then it follows directly that the limit cycles must be alternatively stable and unstable.

A rough observation deriving directly from the Poincaré-Bendixson theorem is also that there are only two kinds of attractors for a two-dimensional flow: (1) stable fixed points (as classified in Annex 6A) and (2) periodic solutions or limit cycles. Consequently, the chaotic motion, which is associated with the so-called **strange attractor**, as we will extensively show in the next subsection, is not possible for two-dimensional flows (see also Lichtenberg and Lieberman, 1983).

Then, it should be noted that the change from a fixed point into a limit cycle (if a single parameter varies), is a phenomenon known as **Hopf bifurcation**. In particular the Hopf-bifurcation theorem (see Guckenheimer and Holmes, 1983, and Marsden and McCracken, 1976) discusses the existence of closed orbits in a neighbourhood of an equilibrium (related to an at least two-dimensional system), if the Jacobian J of the system at hand has a pair of purely imaginary eigenvalues and no other eigenvalues with zero real parts (i.e., tr J = 0 and det J > 0). There are two types of Hopf bifurcation. The bifurcation is '**supercritical**' if each point loses its stability by expelling a stable periodic orbit. It is '**subcritical**' if each point loses its stability by absorbing a non-stable periodic orbit (see Figure 6.3).

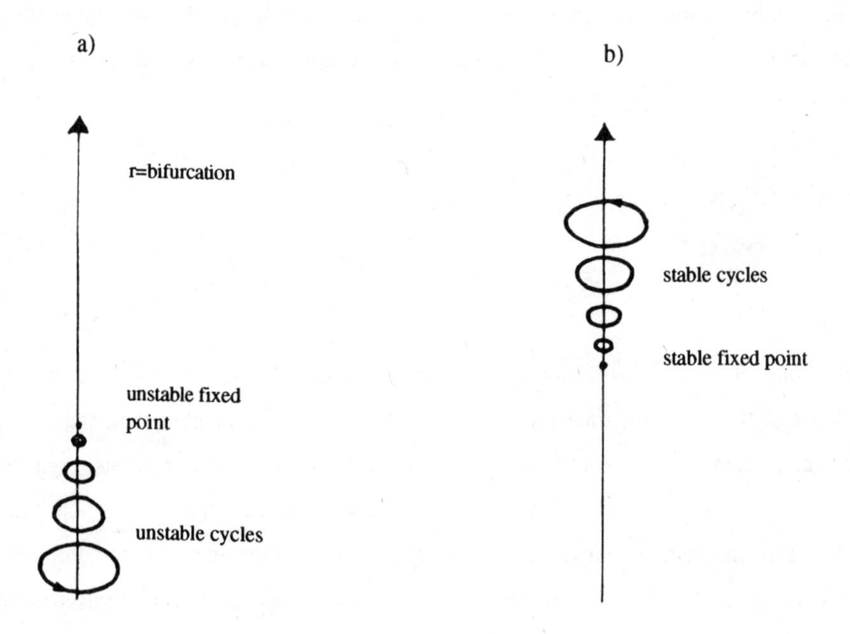

Figure 6.3 Hopf Bifurcation: a) Subcritical, b) Supercritical

It is interesting to note that in an at least three-dimensional system the subcritical bifurcation can be associated with chaotic behaviour (see Marsden and McCracken, 1976), since Newhouse et al. (1978) have shown that after three Hopf bifurcations regular motion becomes highly unstable in favour of motion to a strange attractor (defined here as a bounded region of phase space in which initially close trajectories separate exponentially such that the motion becomes chaotic).

After this brief review of the historical-methodological basis of non-linear dynamics leading, as a final step, to the discovery and analysis of chaos, we will discuss in the next section some key issues in chaos theory with particular reference to the historical background as well as to the 'mathematical' routes leading to chaos.

6.2.2 Key issues in the theory of chaos

Chaos theory has recently attracted widespread interest in a wide range of disciplines. It is actually regarded as a major discovery with a high significance for both the natural and social sciences. Substantial scientific contributions to the theory of chaos can be found in two volumes collecting many basic papers on this issue, viz. Cvitanovic (1984) and Hao (1984). For further details on the mathematics of chaos and non-linearity the reader is referred to some standard books, such as Collet and Eckmann (1980), Devaney (1986), Guckenheimer and Holmes (1983), Hao (1989), Holden (1986), Iooss (1979), Lorenz (1989), Peters (1988), Poston and Stewart (1986), and Schuster (1988). Various informative reviews of chaos theory with reference to its relevance for the social sciences (including economics) can be found in among others, Andersen (1988), Baumol and Benhabib (1989), Benhabib and Day (1981, 1982), Boldrin (1988), Chen (1988), Crilly et al., (1991), Devaney (1986), Guckenheimer and Holmes (1983), Gylfason (1991), Lasota and Mackey (1985), Lucking (1991), Lung (1988), Pohjola (1981), Prigogine and Stengers (1985), Rosser (1991), Stewart (1989), and Stutzer (1980).

The interesting feature of chaos theory is that it is essentially concerned with **deterministic, non-linear dynamic** systems which are able to produce **complex motions** of such a nature that they are sometimes seemingly random. In particular, they incorporate the feature that small uncertainties may grow exponentially (although all time paths are bounded), leading to a broad spectrum of different trajectories in the long run, so that precise or plausible predictions are - under certain conditions - very unlikely, given the high sensitivity on initial conditions.

It is interesting to note here that already at the end of the past century Poincaré (1892) argued that some regulated mechanical systems would develop chaotic behaviour (against the strict determinism claimed by Laplace). "It may happen that small differences in the initial conditions produce very great ones in the final phenomena. A small error in the former will produce an enormous error in the latter. Prediction becomes impossible, and we have the fortuitous phenomenon" (see Zhang, 1991, p. 123).

Consequently, the new logic which has emerged in the area of non-linear dynamics by the introduction of the theory of chaos shows also an interesting psychological appeal; model builders need not necessarily be blamed any more for false predictions, as errors in predictions may be a result of the system's complexity, as can be demonstrated by examining more carefully the properties of the underlying non-linear dynamic model structure.

It is also important here to underline once more that a very important characteristic of non-linear models which can generate chaotic evolutions is that such models exhibit strong sensitivity to initial conditions. Points which are initially close may on average diverge exponentially over time, although their time path is bounded and they may be from time to time briefly very close to each other. Thus, even if we knew the underlying structure exactly, our evaluation of the current state of the system is subject to measurement error and hence it is impossible to predict with sufficient confidence beyond the short run. Similarly, if we knew the current state with perfect precision, but the underlying structure only approximately, the future evolution of the system would also be unpredictable in the long run. The equivalence of the two situations has been demonstrated, among others, by Crutchfield et al. (1982), so that it is difficult to circumnavigate the presence of chaos in various cases.

It is also noteworthy that various scientists appear to use different definitions and formal representations of chaotic behaviour, and - given this lack of unambiguity - it may be worth describing in more detail the origins and the historical evolution of the concept of 'chaos', whose history covers already a few decades.

One of the first examples of a chaos phenomenon was found in the behaviour of deterministic non-periodic flows, which was investigated for the first time by Lorenz (1963), who studied the instability of such flows for forced dissipative hydrodynamical systems and was able to derive the numerical solutions for these systems by means of the convection equations of Saltzman (see Annex 6B). In particular, he found that the projections of the solution trajectories of such dynamic systems followed two spirals - around two steady states - at different surfaces, so that "it is possible for the trajectory to pass back and forth from one spiral to the other without intersecting itself" (Lorenz, 1963, p. 138).

Later on, in subsequent discussions on the concept of chaos, the related term 'strange attractors' was introduced, first by Ruelle and Takens (1971), in order to indicate "an exponential separation of orbits" (see Eckmann and Ruelle, 1985, p. 619). In fact, the Lorenz model may be regarded as the first example in the scientific literature having such strange attractors. The Lorenz model is illustrated in Annex 6B.

In a later stage, Li and Yorke (1975) christened 'chaotic' a system with strange attractors or a dynamic situation exhibiting aperiodic - though bounded - trajectories. In this context the well-known Li and Yorke theorem (1975) 'Period three implies chaos' is fundamental here, since it shows the existence of chaotic motion in simple first order differential equations of the following type:

$$x_{t+1} = a\, x_t\, (1-x_t) \tag{6.2}$$

This model has been extensively discussed by May (1976) for the description of biological population growth x. The parameter a is the growth parameter reflecting the maximum per capita rate of increase of the time-dependent variable x. The logistic map (6.1) requires for its existence that $0 \leq x \leq 1$ and $0<a<4$. Equation (6.1) exhibits fixed points (or equilibrium values, i.e., values that do not change when the mapping is iterated) as well as bifurcations of fixed points (see Figure 6.4).

In particular, for $a>3.824..$ a cycle of period 3 appears (e.g., in biology a population value which reiterates every third generation), beyond which there are cycles in every integer period, as well as an uncountable number of aperiodic trajectories; in other words, according to the above mentioned statement of Li and Yorke, this is a typical example of a chaotic region.

It is noteworthy here that often - instead of the term 'chaos' - alternative expressions with the same meaning are used by various authors, such as '**dynamical stochasticity**', '**self-generated noise**', or '**intrinsic stochasticity**' (see Hao, 1984).

After these introductory remarks, we now need a more strict definition of chaos. Ott (1981) and Pacini (1986) define a system as chaotic, if there exists an uncountable, invariant set A of initial conditions such that all trajectories starting in A meet the following requirements:

- they never repeat themselves (aperiodic elements of A);

- they neither attract nor are attracted by other trajectories;
- they show a sensitive dependence on initial conditions.

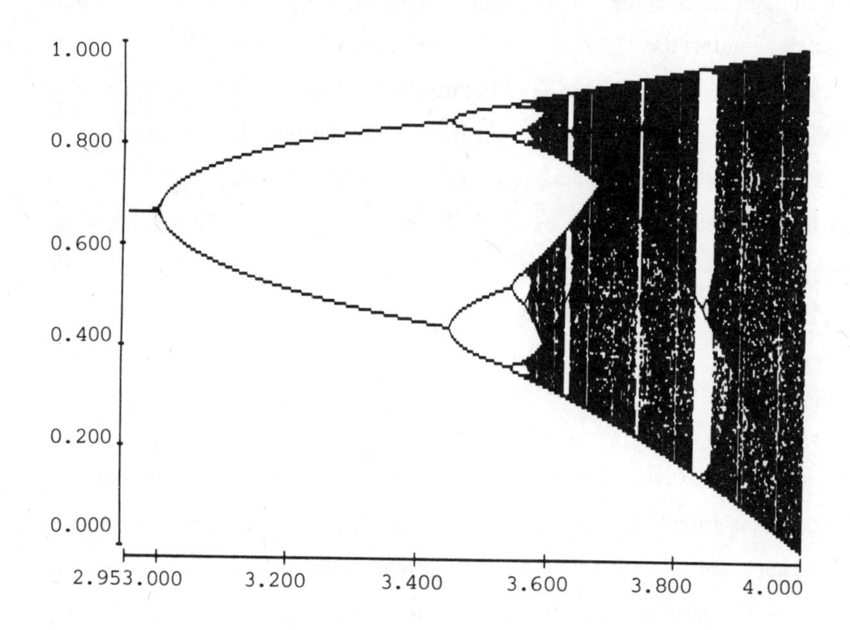

Figure 6.4 Bifurcation Diagrams of a May Logistic Map (2.9<a<3.9)

Moreover, together with chaotic trajectories, periodic points of every order coexist, although most of them are not acting as an attractor. Thus the resulting aperiodic and interlaced cycles are able to produce very complicated forms leading the above mentioned 'strange' attractor.

From a mathematical point of view there are at least three routes to chaos by which non-linear systems can become chaotic (see Schuster, 1988). The first route is known as '**Feigenbaum route**'. In particular Feigenbaum (1978) showed that the route to chaos that is found in the logistic map (6.2) occurs in all first-order difference equations:

$$x_{t+1} = f_a(x_t) \tag{6.3}$$

in which the parameter-valued function $f_a(x_t)$ has only a single maximum in the unit interval $0 \leq x_t \leq 1$. He also discovered that the behaviour at the transition to chaos is governed by two universal scaling parameters (the so-called Feigenbaum constants). In particular Feigenbaum found that the number of fixed points of $f_a(x)$ doubles at distinct, increasing values of the parameter a. Moreover, a value $a = a^\infty$ exists at which the number of fixed points become infinite; beyond this value the behaviour of the iterates becomes irregular (chaotic).

The second route to chaos, known as the **'intermittency route'**, has been discovered by Manneville and Pomeau (1979). Intermittency means the occurrence of a signal which alternates randomly between long regular phases (intermissions) and relatively short irregular bursts (intermittent bursts). It has been shown, by analyzing the differential equations of the Lorenz model, that the average number of these bursts increases with an external parameter until the motion becomes truly chaotic.

The third route was found by Ruelle and Takens (1971) and Newhouse et al. (1978). It is known as the **'Ruelle-Takens-Newhouse route'**. These authors showed in particular that after three Hopf bifurcations, it is 'likely' that regular motion becomes highly unstable in favour of motion to a strange attractor (see Figure 6.5 where ω represents the frequency of the related orbits).

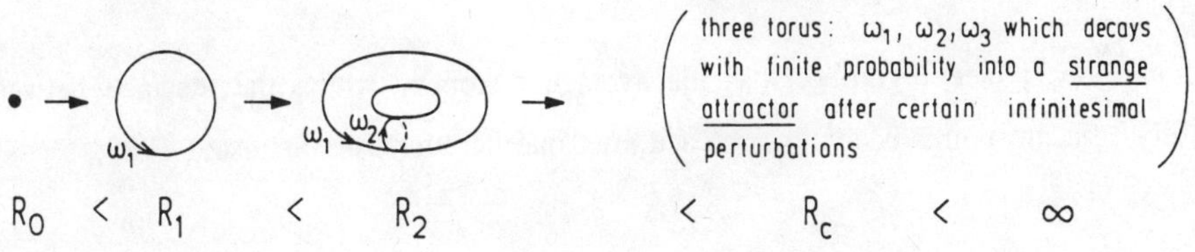

ω_1 ω_1 ω_2

three torus: $\omega_1, \omega_2, \omega_3$ which decays with finite probability into a strange attractor after certain infinitesimal perturbations

$R_0 < R_1 < R_2 < R_c < \infty$

Figure 6.5 The Ruelle-Takens-Newhouse Route to Chaos.
Source: Schuster (1988, p. 148)

It should be noted that, provided the dimension of the system is large enough, according to the sequences of Hopf bifurcations, the limit cycle bifurcates into a so-called **torus**. In Figure 6.6 a two-dimensional torus T^2, constructed by an appropriate gluing of a two-dimensional area, is illustrated. Further bifurcations may lead to the tori T^3, T^4, etc.

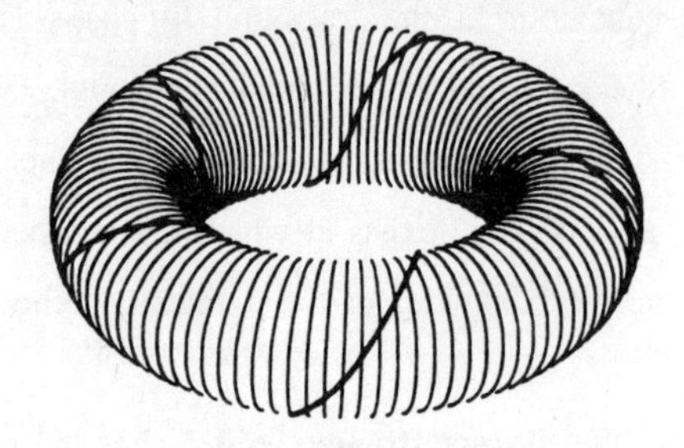

Figure 6.6 The Two-dimensional Torus T^2
Source: Haken (1983b, p. 29)

Another important way of showing the existence of chaos is provided by the **Liapunov exponents** which measure the exponential divergence of nearby trajectories, characterizing the chaotic motion (as already mentioned). The Liapunov exponents $\lambda(x_o)$ are analytically defined (e.g., in Haken, 1983a, and Schuster, 1988) by considering the general discrete map:

$$x_{t+1} = f(x_t) \tag{6.4}$$

and $\exp[\lambda(x_o)]$ (see Figure 6.7) as the average factor by which the distance between closely adjacent points becomes stretched after one iteration (see Schuster, 1988):

$$\underset{x_0 \qquad x_0 + \xi}{\overline{\qquad \xi \qquad}} \quad \overset{\text{n iterations}}{\Longrightarrow} \quad \underset{f^n(x_0) \qquad\qquad f^n(_{x0} + \xi)}{\overline{\qquad \xi \exp [n\lambda(x_0)] \qquad}}$$

Figure 6.7 Definition of the Liapunov Exponent

In a formal way we can write:

$$\xi \exp[n\lambda(x_0)] = | f^n(x_0 + \xi) - f^n(x_0)| \tag{6.5}$$

which leads to the following final expression (see Schuster, 1988):

$$\lambda(x_o) = \lim_{n \longrightarrow \infty} \frac{1}{n} \ln \left| \frac{df^n(x_o)}{dx_o} \right| \tag{6.6}$$

Consequently, at least one Liapunov exponent has to be positive in a chaotic attractor, since in this case neighbouring trajectories depart very quickly from each other. Obviously Liapunov exponents can also be defined for differential equations (see Zhang, 1990). Thus in one dimension the Liapunov exponents are negative, since there are only stable fixed points; also in two dimensions there are not positive Liapunov exponents, since the possible classes of attractors are only two: stable fixed points and limit cycles (see Section 6.2.1). In particular for a limit cycle one Liapunov exponent is zero. Summarizing, we can see the most typical cases in Figure 6.8.

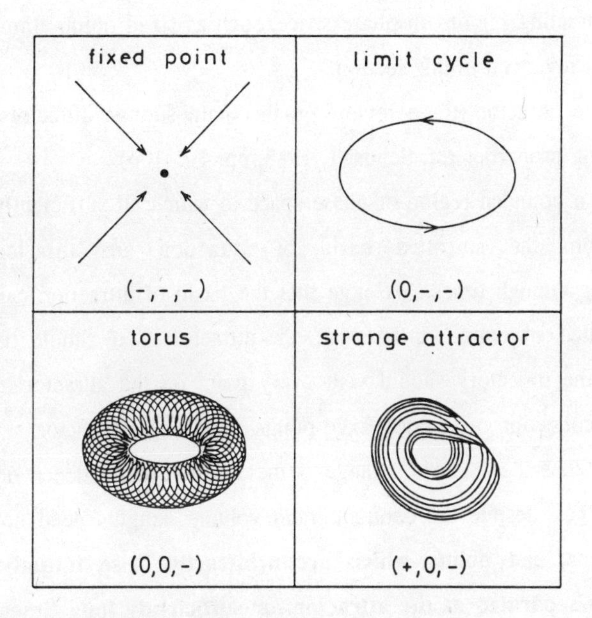

Figure 6.8 Connections between the Attractors Embedded in Three-Dimensional Space and Their Three Liapunov Exponents (in brackets)
Source: Schuster (1988, p. 139)

It is interesting to note that Liapunov exponents are also related to the loss of information about the state of the system. Consequently, the Liapunov exponent is also connected with Kolmogorov entropy K (see Shannon and Weaver, 1949) which measures, according to information theory (see also Chapter 2), our ignorance about the system. In other words, K is proportional to the rate at which information about the state of the dynamic system is lost in the course of time by following Shannon's formulation (see equation (2.47)). Thus K is positive if the system displays deterministic chaos, and zero for regular motion. Moreover K is connected with Liapunov exponents since it is equal to the (average) sum of positive Liapunov exponents (see Schuster, 1988).

Three common approaches to measuring instability - via Liapunov exponents - are also found in the Eckmann and Ruelle (1985), Kurths and Herzel (1987) and the Wolf et al. algorithms (1985) (see Sayers, 1991).

It should be noted that the analysis carried out in this subsection is related to dissipative systems for which volume elements in phase space shrink with increasing time. This property leads to attracting regions in phase space, such as fixed points, limit cycles and strange attractors, as reviewed in this section.

Summarizing, a strange attractor (for a review on the main strange attractors, see Annex 6B) has the following properties (cf. Schuster, 1988, pp. 105-106):

a) "It is an attractor, i.e., a bounded region of phase space to which all sufficiently close trajectories from the so-called basin of attraction are attracted asymptotically for long enough times. We note that the basin of attraction can have a very complicated structure. Furthermore, the attractor itself should be indecomposable; i.e., the trajectory should visit every point on the attractor in the course of time. A collection of isolated fixed points is no single attractor.

b) The property which makes the attractor strange is the sensitive dependence on the initial conditions; i.e., despite the contraction in volume, lengths need not shrink in all directions, and **points which are arbitrarily close initially, become exponentially separated at the attractor for sufficiently long times**. This leads to a positive Kolmogorov entropy."

It is interesting to note that also in conservative systems where the volume is conserved one can find chaos (with a positive Kolmogorov entropy); in other words there are strange or chaotic regions in phase space, but they are not attractive and can be densely interwoven with regular regions (see Schuster, 1988).

A final remark concerns the 'strong' connection between strange attractors and fractal (i.e., non-integer) dimensions (see, e.g., Annex 6B.1). However, even though chaos and fractal geometry are closely related, the two fields have not been incorporated so far into a unified framework (see Rosser, 1991, and Vicsek, 1989).

In conclusion, the discovery of 'chaos' seems to have created a new paradigm in scientific modelling which generates intriguing research questions. Firstly, the process of verifying theories on dynamic systems behaviour through conventional predictions becomes more problematic in case of chaotic systems (see Ornstein and Weiss, 1991). And secondly, the concept of chaos demonstrates that a system can have a complex global behaviour at large which in general cannot be deduced from knowledge of its constituent parts. Thus "chaos provides a mechanism that allows for free will within a world governed by deterministic laws" (see Crutchfield et al., 1986, p. 57).

6.3 Spatial Applications of Chaos Theory: A Brief Survey

6.3.1 Introduction

In the previous subsections we mentioned that the first interesting studies on chaotic features of complex systems were carried out in physics, meteorology and natural sciences. In recent years also the economics discipline has witnessed an increasing wave of contributions in the use of chaos theory for analyzing economic-spatial patterns. The main purpose of the use of this theory in the social sciences in general was to obtain better insight into the underlying causes of unforeseeable evolutionary patterns of complex dynamic systems.

It is interesting to mention here a sample of the numerous applications and illustrations of chaos theory in the following fields of economics:

- growth and business cycle theory (Balducci et al., 1984, Benhabib and Day, 1982, Boldrin, 1988, Casti, 1979, Day, 1982, Funke, 1987, Guckenheimer et al., 1977, Grandmont, 1985, 1986, Hommes et al., 1990, Scheinkman and LeBaron, 1989, and Stutzer, 1980)
- cobweb models (Chiarella, 1988)
- long waves analysis (Nijkamp, 1987a, Rasmussen et al., 1985, and Sterman, 1985)

- R&D analysis (Baumol and Wolff, 1983, and Nijkamp et al., 1991)
- consumer behaviour (Benhabib and Day, 1981)
- duopoly theory (Rand, 1978, and Dana and Montrucchio, 1986)
- economic competition (Deneckere and Pelikan, 1986, and Ricci, 1985)
- international trade (Lorenz, 1987)
- competitive interactions between individual firms (Albin, 1987)
- equilibrium theory (Hommes and Nusse, 1989, Nusse and Hommes, 1990, and Radzicki, 1990)
- management systems (Rasmussen and Mosekilde, 1988)
- monetary analysis of the exchange rate (De Grauwe and Dewachter, 1990).

Unfortunately, it is disappointing to observe that most applications of chaos theory in economics (and in general the social sciences) lack empirical content. Brock (1989) claims that as yet no class of structural economic models has been **estimated** which allows for chaotic behaviour **and** in which the estimated model parameters are indeed in the chaotic range. Moreover, statistical tests (such as the Brock-Dechert-Scheinkman (1987) statistic, or the largest Liapunov exponent) which have been designed to detect chaos in time series without **a priori** specification of the nature of the data generating process, have not provided as yet unambiguous empirical support for the presence of chaos in observable economic processes.

This does not imply that non-linear dynamic structure is absent from economic and financial time series but the available tests are not able to identify the nature of this structure. For example, Brock and Sayers (1988) found evidence of nonlinearity in the following US statistics: employment and unemployment (quarterly), industrial production (monthly) and pigiron production (annually). Empirical evidence of nonlinear dynamics is now also emerging elsewhere, e.g. in weekly price observations in German agricultural markets (Finkenstädt and Kuhbier, 1990) and Austrian demographic data (Prskawetz, 1990). However, non-linear determinism tends to be absent in many macro-economic aggregates such as GNP and private investment. One rare finding of chaos in monetary aggregates was recorded by Barnett and Chen (1988), but their finding has been convincingly challenged by Ramsey et al. (1990).

Two interesting reasons why linear modelling of economic time series may be adequate and why a non-linear deterministic structure may therefore be absent or undetectable in such cases are the following (see Nijkamp and Poot, 1991). Firstly, it can be easily demonstrated that deterministic chaos requires **positive feedback** loops which may be found in phenomena such as industrial clustering, networking and the growth of cities, but which might less likely occur in financial and economic time series (Brock, 1989). Secondly, another reason for the difficulty in detecting chaos is that economic time series are, after detrending, **inherently noisy** due to measurement errors and outside shocks. In this case there may be some deterministic structure underlying the stationary fluctuations, but the high-dimensional chaos generated by this process may be indistinguishable from true randomness.

A final remark concerns some criticism on the theoretical economic models displaying chaotic behaviour. First it should be noted that the underlying hypothesis of these models is that aggregate fluctuations might represent an endogenous phenomenon that would persist even in the absence of stochastic 'shocks' in the economy. Clearly, this usual interpretation has to be judged with caution, since chaos is **only** a possibility which is not bound to happen. Second, models generating chaotic behaviour include so far only a very limited number of equations, whereas in reality economic systems are usually much more complicated. And finally in many cases the economic justification underlying the specification of these chaotic models is not always clear. It seems thus evident that in economics chaos is not a necessity but only a possibility, so that it is at least a valid research endeavour to specify models allowing for chaos even if any testable implication is difficult to carry out.

After these introductory considerations we will now address our efforts towards one specific research area, viz., regional economics. The issue of spatial dynamics has drawn quite some attention in the recent past; consequently we will present a selection of applications displaying chaotic behaviour in **spatial** systems, and also touch upon some theoretical aspects.

6.3.2 Dendrinos

Dendrinos has explored chaotic dynamics (mostly in socio-spatial systems) from both a theoretical and an empirical perspective. In a first article (Dendrinos, 1984a), he uses a May-type differential equation for modelling urban macro dynamics. More specifically he adopts the following form:

$$y(t+1) = A\, y(t)[B-y(t)] \tag{6.7}$$

where $y(t) \leq B$ represents the population and $A, B > 0$ are relevant parameters. The author shows that formulation (6.7) satisfactorily replicates **urban aggregate** dynamics in the U.S. for the period 1890-1980. He observes in particular that the size of urban areas always affects (inversely) the amplitude or the number of the oscillations required to reach a steady state.

Furthermore, he demonstrates that the actual (A, B) values associated with U.S. cities are - in terms of the difference between the current parameters versus the values corresponding to turbulent behaviour - spatially close to the turbulent regime in the (A, B) space, but - in terms of speed of change of the parameters - over time remote from it. In a second contribution (Dendrinos, 1986) the author tries to overcome the problem of the choice between discrete and continuous dynamics by connecting spatial flows to continuous fluid convection dynamics on the basis of earlier research undertaken by Lorenz (1963) and Sparrow (1982). He then applies a Lorenz system (described in Annex 6B.3) to regional employment by adopting seven (which can in fact be reduced to four) - instead of three - parameters which obviously produce less efficient (i.e., less chaotic) outcomes than the Lorenz model. The results of this model show that the trajectories converge towards periodic orbits, although such orbits may not always be well defined. Moreover, the origin of this system might be an attractor - instead of a repulsor like in the Lorenz model - implying an extinction of the socio-spatial system at hand.

The conclusion seems that unexpected behaviour may exist depending on fluctuations in the model parameters induced from exogenous changes.

Then, in a more recent contribution (Dendrinos, 1988), Dendrinos explores deterministic turbulent transport developments. In particular, gravitational spatial interaction is incorporated into a Volterra-Lotka spatial population dynamics model. It is shown that the amplitude of the ecologically driven stock dynamics constrains the magnitude of oscillation in the dynamics of spatial gravitational flow interactions. Furthermore, the author considers a spatial interaction problem (i.e., a congested transportation framework) by showing that the discrete iterative dynamic equilibria are either stable attractors or stable two-period cycles. Finally, in his last contribution, Dendrinos (1991) finds that instability is the central ingredient of macro urban evolution at the global scale.

6.3.3 Dendrinos and Sonis

Dendrinos and Sonis (1988) have investigated socio-spatial dynamics on the basis of a one-dimensional discrete map (we recall here that a discrete map is a mathematical relationship that allows the next value X_{t+1} of a quantity X to be obtained from its present value X_t). The authors studied discrete **regional relative** population dynamics by following the line of research described in Section 6.3.2 (Dendrinos, 1984a). Their analysis shows the importance of the level of spatial disaggregation for the analysis of dynamic instability; in particular, they show that for the U.S. population the qualitative dynamics of the U.S. regional paths differ significantly when different levels of spatial disaggregation are employed.

In another article (Dendrinos and Sonis, 1987) the authors explore the onset (i.e., the initial point) of turbulence in discrete relative multiple spatial dynamics by demonstrating - for a three location problem - local (when two variables out of three are in an oscillatory motion, in particular in a stable two-period cycle) and partial (when the cycles are periodic) turbulence. Furthermore, they also show (see Sonis and Dendrinos, 1987) that the well-known Feigenbaum sequence (i.e., the well-known rule-sequence of bifurcation values for which a period doubling occurs) does not hold over the bifurcation parameter sequence for period-doubling cycles related to more complex maps than the usual logistic one. The universality of the log-linear relative dynamics map is then studied in detail, with particular reference to the case of one population - three locations log-linear relative dynamics (Sonis, 1990). In particular Sonis has studied the existence and number of non-period fixed points for this model.

6.3.4 Mosekilde, Aracil and Allen

Along the research direction developed by Reiner et al. (see next Section 6.3.6), Mosekilde et al. (1988) used a similar migratory model for illustrating the concepts of attractors, bifurcations, Poincaré sections (i.e., cross sections of the orbits of one dimension lower than the space of the orbits) and return maps.

Then in a further contribution (Sturis and Mosekilde, 1988) Poincaré sections are used for showing the existence of strange attractors in this four-dimensional migratory system. The inclination to move is utilized as a bifurcation parameter.

6.3.5 Nijkamp

Nijkamp (1987a) has developed a simple model for analyzing endogenous long-term spatial fluctuations. On the basis of a dynamic Cobb-Douglas production function he is able to derive the following dynamic relationship:

$$Dy_t = \tilde{y}_t(y^{max} - ky_{t-1})\, y_{t-1} / y^{max} \tag{6.8}$$

where $Dy_t = y_t - y_{t-1}$; y_t represents the regional share in the national production; $\tilde{y}_t$ is the rate of change in the original quasi-production function (incorporating also infrastructure capital and R&D capital) and y^{max} reflects the maximum capacity level, beyond which congestion factors lead to a negative marginal product.

It is evident that (6.8) is essentially a May-type model, so that (6.8) is able to generate a wide variety of dynamic growth patterns, although in principle the behaviour of such a model is determined by the initial conditions of the system and by its growth rate.

In a subsequent paper (Nijkamp et al., 1991) the previous model has been extended toward a Harrod type growth model by incorporating also investment and savings behaviour. Next, R&D investments are endogenized, by assuming that the growth path of income, consumption and investment is co-determined by R&D investments. By imposing next the condition of a declining marginal efficiency of R&D expenditures and finally even of a saturation level, one faces the possibility of diseconomies of scale. The (maximum) saturation level plays the same role as y^{max} in (6.8). By means of various simulation experiments in both a single region and a multi-region system the authors were able to analyze the dynamic behaviour of an evolutionary spatial economic system.

Then in a final paper (Nijkamp and Poot, 1991) it is found that for plausible parameter values chaotic regimes in model (6.8) are unlikely. However, in case of rapid transitions or sudden adjustments such chaos patterns may temporarily emerge.

6.3.6 Reiner, Munz, Haag and Weidlich

Reiner, Munz, Haag and Weidlich (1986) have shown that migratory systems, modelled according to the stochastic equation of motion (the so-called **master equation**), can "provide another example, to which the concepts of strange attractors and deterministic chaos fully apply under certain trend parameter conditions" (Reiner et al., 1986, p. 305). To this purpose the authors analyze their model by introducing and applying interesting new concepts such as the Liapunov exponents and fractal dimensions. Their conclusion is that an endogenous migratory system exhibits chaotic behaviour only under very strong conditions. Under usual and less strong conditions regular behaviour of migratory systems are likely to emerge.

6.3.7 White

White (1985) has investigated the conditions under which chaotic behaviour arises in an industrial system. In particular, he models the growth (or decline) of each sector in each centre by using difference equations of the following type:

$$X_{ij,t+1} = X_{ij,t} + r_j \, (P_{ij,t}) \tag{6.9}$$

where $X_{ij,t}$ represents the size of sector j in centre i at time t, r_j the intrinsic growth rate of the sector, and $P_{ij,t}$ the profit generated (which is depending on the aggregate amount produced in the sector concerned by all centres). The way the sectoral interactions are specified determines of course to a large extent the evolutionary pattern of this spatial system. It is clear that this system may exhibit both stable and unstable behaviour.

It is evident that - for a certain specification of the $P_{ij,t}$ terms - equation (6.9) belongs to a general family of Verhulst equations (see Peitgen and Richter, 1986) discussed also by May (1976) and Yorke and Yorke (1981) in the framework of chaotic behaviour.

White's simulation results show that the value of r for which chaotic behaviour appears is inversely related to the number of the centres. Furthermore, the author stresses that the onset of chaotic behaviour is not so clear as pointed out by previous authors. In his interesting contribution he investigates in particular different degrees of chaos for the one-equation models.

6.3.8 Zhang

Along the research line developed by Nijkamp and Reggiani (see Section 6.4), Zhang (1991) interprets a Lorenz system in the context of urban dynamics. In particular he considers the following three variables:

$$\begin{aligned} x &= \text{output of the urban system} \\ y &= \text{number of residents} \\ z &= \text{land rent} \end{aligned} \tag{6.10}$$

and then introduces nine parameters. By using transformations of the parameters as well as of the variables, Zhang finally shows the equivalence between his model and the Lorenz model. However, no simulation is carried out also in order to investigate the chaotic (and hence possibly negative values) of the population variables.

6.3.9 Concluding remarks

The previous examples show that much research still has to be undertaken in order to fully understand the chaotic possibilities in spatial economic and geographic systems. Obviously so far the research has mainly been devoted to find theoretical models that predict chaos as a logical outcome of 'reasonable' hypotheses; the lack of available data often hinders researchers to test adequately the specification of chaotic models and to find statistical/econometric evidence of chaotic movements. Then also from a theoretical point of view the mathematical conditions (and hence the range of the parameter values) for reaching chaos are very strict.

In the next section we will illustrate the possibilities of chaotic behaviour by presenting an example in the specific field of urban dynamics. Owing to the lack of data, simulation experiments will therefore be carried out.

6.4 A Model of Chaos for Spatial Interaction and Urban Dynamics

6.4.1 Introduction

In this section a chaotic **macro**-model of urban dynamics will be examined. It should already be noted here that in Chapter 7 the possibility of chaotic motion in **micro** models in relation to discrete choice behaviour will be discussed. Here we will describe in particular three macro-behavioural equations illustrating the evolution of a city characterized during a certain phase of its life cycle by a structural decline. Structural decline means a permanent negative impulse toward extinction: this hypothesis is plausible since empirical evidence on urban decline of many cities can be found in a great number of recent studies on urban evolution (see, e.g., Dendrinos, 1980, and Van den Berg et al., 1987).

The three key variables assumed in our dynamic model are:

- city size (measured in terms of number of inhabitants)
- employment potential (measured in terms of employment rate, i.e., the share of working population in the total population).
- urban attractiveness (measured in terms of immigration rate, i.e., the share of immigrants in the total population)

The latter two variables can exert a negative impulse in case of unemployment or in case of urban repulsion effects, respectively.

The first assumption is based on the idea that in a structurally declining area the growth rate (σ_2) of the population is negative, although this may be compensated by a rise in the employment rate through the parameter (σ_1), i.e.,:

$$\dot{x} = \sigma_1 y - \sigma_2 x \tag{6.11}$$

where x and y represent population size and employment rate, respectively; σ_1 and σ_2 denote the related growth and decline rates.

The same type of relationship is also assumed for the urban attractiveness, so that:

$$\dot{z} = \tau y - \beta_1 z \tag{6.12}$$

where z is the immigration rate with a growth rate β_1; τ is the growth rate of the employment rate.

Then, the growth rate τ is assumed to be positively correlated (through the parameter β_2) with agglomeration economies emerging from city size, i.e.,:

$$\tau = \beta_2 \, x \tag{6.13}$$

Therefore, we get for (6.12) the following final expression:

$$\dot{y} = - \delta_1 \, y - \phi z + \delta_3 \, x \tag{6.14}$$

The last assumption in our dynamic model is related to the evolution of the employment potential. In particular we assume that the employment potential (in terms of employed vis-à-vis unemployed persons) of a structurally declining city has a negative growth rate ($-\delta_1$), which may be reinforced by a negative immigration rate ($-\phi$), but which may also be positively influenced by a rise in city size through the parameter δ_3, i.e.,:

$$\dot{z} = - \beta_1 \, z + \beta_2 \, xy \tag{6.15}$$

Furthermore, we assume that the parameter ϕ in (6.15) is positively correlated to the city size through the parameter δ_2, i.e.,:

$$\phi = \delta_2 \, x \tag{6.16}$$

so that the expression (6.16) becomes:

$$\dot{y} = - \delta_1 \, y - \delta_2 \, xz + \delta_3 \, x \tag{6.17}$$

It is easy to see that the dynamic equations (6.11), (6.14) and (6.17) are essentially Lorenz equations (see Annex 6A.3) with seven parameters. The stability conditions of this system are shown in Annex 6C, where also the conditions for the onset of a Hopf bifurcation are illustrated. On the basis of this analysis various simulation experiments will then be carried out in order to visualize the long-term behaviour of our dynamic model, by varying the parameter values. The related results are reported in Section 6.4.2.

It is interesting to note that if we want to control the Lorenz model by means of an optimization process (e.g., by means of optimal control theory), we may obtain optimal trajectories which are again oscillating (see Nijkamp and Reggiani, 1990c). This outlines the importance of the parameter values; this also implies that policy makers may try to influence the parameters whose critical values determine the oscillating behaviour of the model.

6.4.2 Results of simulation experiments

In this section the results of various simulation experiments will be discussed. In particular we will present a set of interesting examples which show the various possibilities of urban evolution (regular/irregular) depending on the values of the parameters and initial conditions.

6.4.2.1 The onset of chaotic motion

For this simulation the following parameter values will be assumed:

$\sigma_1 = 0.1$ $\sigma_2 = 0.001$

$\delta_1 = 0.1$ $\delta_2 = 0.005$ $\delta_3 = 0.001$

$\beta_1 = 0.01$ $\beta_2 = 0.0001$

with the following initial values:

$x = 100$ $y = 0.5$ $z = 0.1$

In this example we have chosen the critical condition $\sigma_1\delta_3 = \sigma_2\delta_1$ after which unstable motion begins (as illustrated in Annex 6C). However, we are still far from the conditions necessary to get a Hopf bifurcation, with a likely chaotic behaviour.

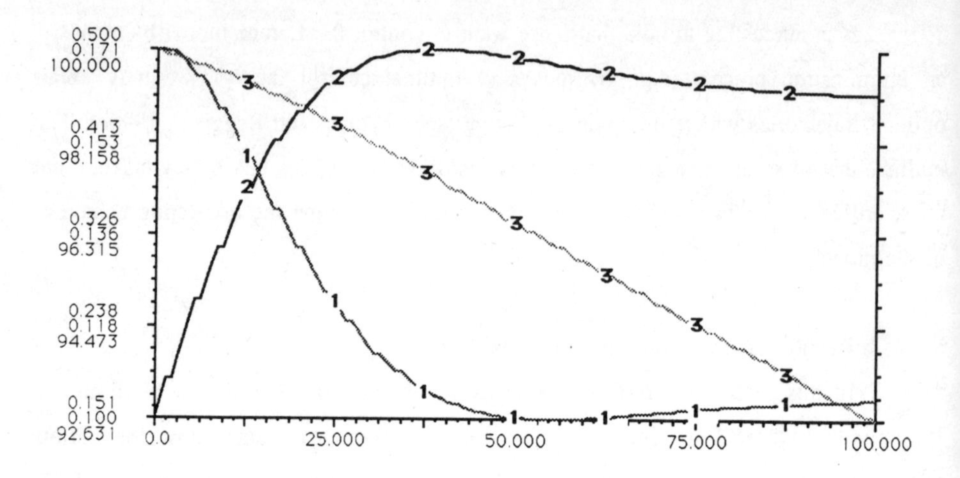

Figure 6.9 Modest Urban Decline
y-axis = 1:employment rate; 2: immigration rate; 3: population (in descending order from the top)
x-axis = time

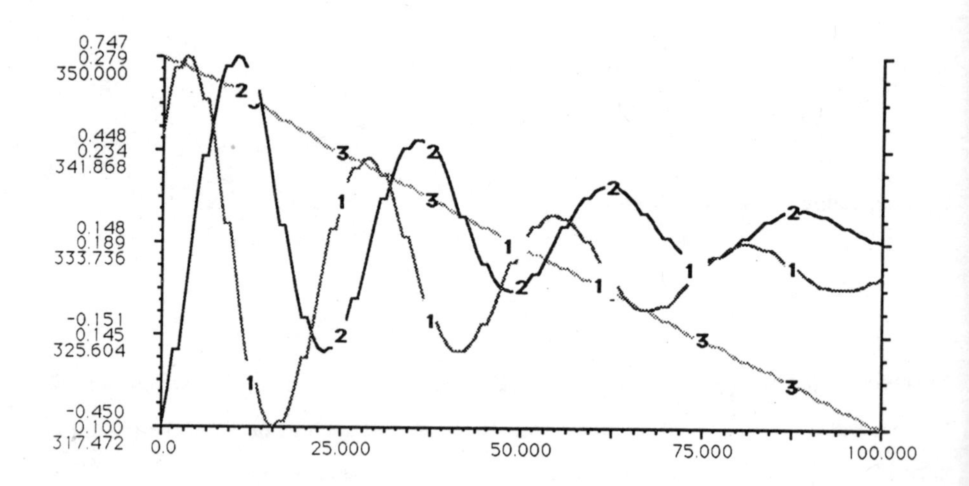

Figure 6.10 Modest Urban Decline with Fluctuations
y-axis = 1: employment rate; 2: immigration rate; 3: population (in descending order from the top)
x-axis = time

Figure 6.9 shows a motion of modest decline (without oscillations) for the urban spatial interaction system concerned. In particular we can see a gradual decline for the total population, a clear decreasing pattern for the employment rate and a contrasting growth pattern for the immigration rate. This contrasting behaviour shows that the inflow increases until the critical value of the minimum employment rate is reached; after that also a slight decline in the immigration rate takes place.

It is evident that this result strongly depends on initial conditions. Other simulation experiments show, for example, the emergence of fluctuations by increasing the initial values of population or by increasing the value of β_2.

For example, we can see in Figure 6.10 (where we have assumed as initial condition x = 350, while keeping the other parameters constant) oscillating paths especially for the immigration rate and employment rate. On the other hand, if we increase the value of β_2, so that it approaches the value 1 like in the standard Lorenz system, we observe the emergence of fluctuations in all variables (see Figure 6.11).

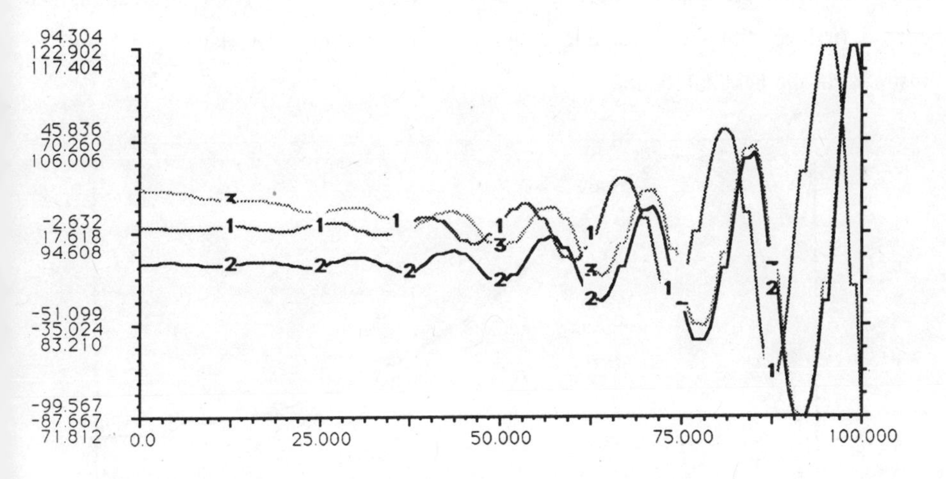

Figure 6.11 The Onset of Fluctuations
y-axis = 1: employment rate; 2: immigration rate; 3: population (in descending order from the top)
x-axis = time

It should be noted that the result presented in Figure 6.11, in comparison with the one shown in Figure 6.9, is obtained only by varying $\beta_2 = 0.005$. However, if we decrease the initial value of the population (e.g., by imposing x = 10), we obtain again a modest regular urban decline, despite the increase of β_2 (see Figure 6.12).

Obviously, other ranges of parameter values can determine the onset of turbulence, by keeping the condition $\sigma_1\delta_3 = \sigma_2\delta_1$. An example is shown in Figure 6.13, where we have assumed the following parameter values with the usual initial conditions (x = 100; y = 0.5; z = 0.1):

$\sigma_1 = 0.1$ $\sigma_2 = 0.01$

$\delta_1 = 0.1$ $\delta_2 = 0.01$ $\delta_3 = 0.01$

$\beta_1 = 0.1$ $\beta_2 = 0.005$

It is evident from the previous examples that the unstable motion related to our system absolutely depends on the initial conditions and the parameter values, even if we are still far from the mathematical conditions necessary to get a Hopf bifurcation (and hence a first condition for reaching a likely chaotic motion). The latter point will be illustrated in the next subsection.

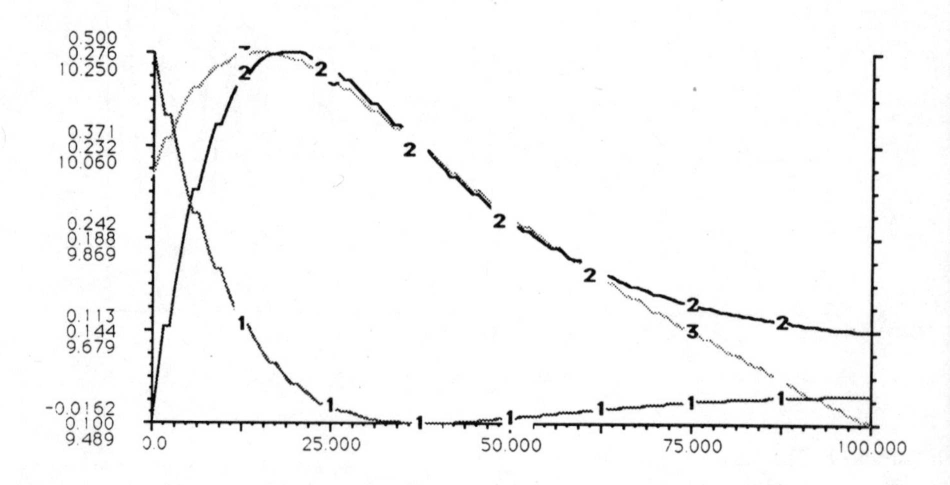

Figure 6.12 Regular Urban Decline
y-axis = 1: employment rate; 2: immigration rate; 3: population (in descending order from the top)
x-axis = time

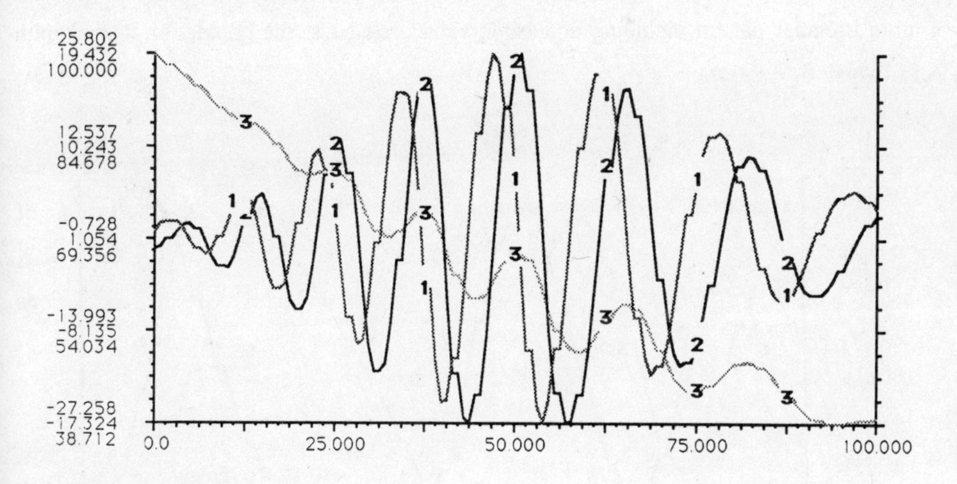

Figure 6.13 Urban Decline with Fluctuations
y-axis = 1: employment rate; 2: immigration rate; 3: population (in descending order from the top)
x-axis = time

6.4.2.2 Chaotic urban evolution

Here we consider a range of parameter values satisfying the conditions of a Hopf bifurcation (see Annex 6C). Since we are likely to approach a chaotic behaviour from a mathematical point of view (see again Annex 6C), we use the word 'chaotic' in this sense. The parameters adopted in this example are the following:

$\sigma_1 = 0.1$ $\quad$ $\sigma_2 = 0.015$

$\delta_1 = 0.01$ $\quad$ $\delta_2 = 0.01$ $\quad$ $\delta_3 = 0.026$

$\beta_1 = 0.001$ $\quad$ $\beta_2 = 0.001$

The initial values of the variables are the same as those of the previous examples (x = 100; y = 0.5; z = 0.1).

It should be noted that here the parameter δ_3 satisfies the conditions (6C.8) for the occurrence of a Hopf bifurcation. The results are shown in Figure 6.14.

These results show an interesting chaotic decrease in urban population caused by parameter effects, whilst the immigration rates and the employment rates show an increasing fluctuating pattern in the same period. It is clear that by increasing δ_3 we get a more irregular pattern including unfeasible values related to the population (see Figure 6.15, where $\delta_3 = 0.04$).

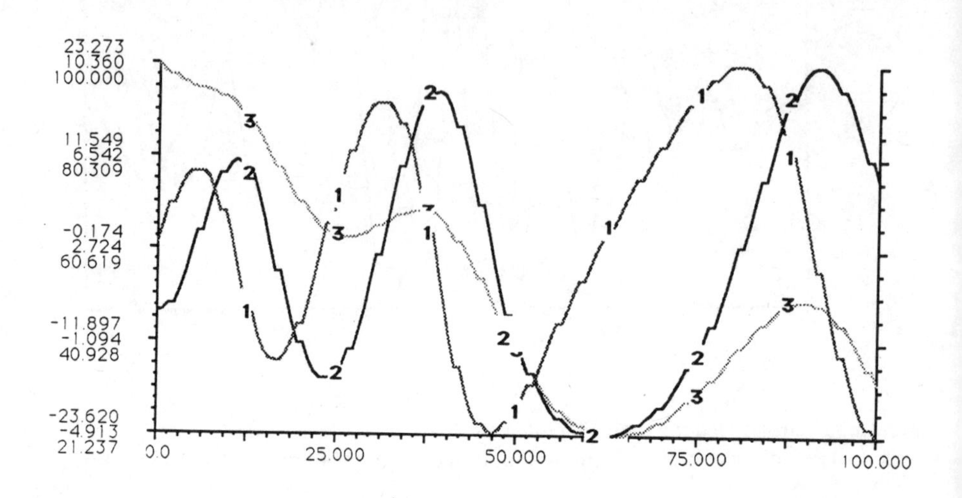

Figure 6.14 Chaotic Urban Decline
y-axis = 1: employment rate; 2: immigration rate; 3: population (in descending order from the top)
x-axis = time

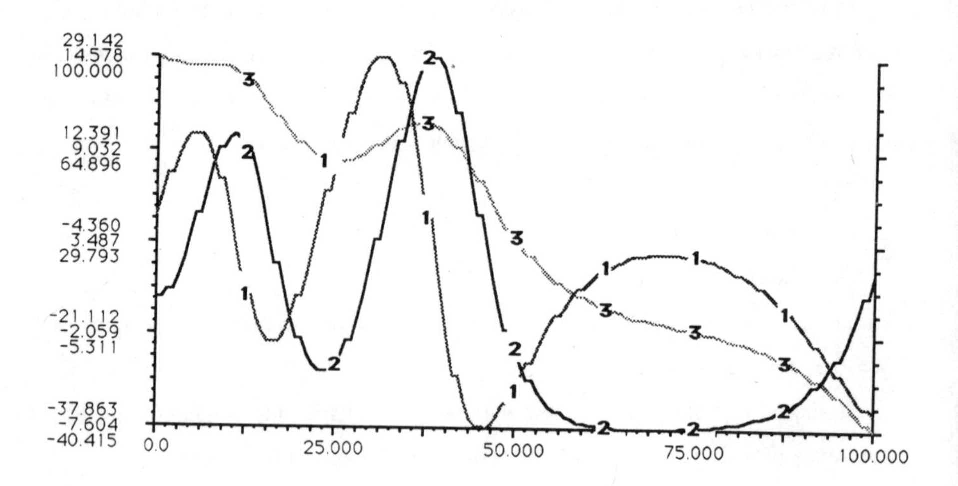

Figure 6.15 Chaotic Urban Decline with Unfeasible Values for Population
y-axis = 1: employment rate; 2: immigration rate; 3: population (in descending order from the top)
x-axis = time

The same type of results may emerge because of a size effect, i.e., by increasing, for example, the initial value of the population (see Figure 6.16 where δ_3 = 0.026 but x = 150).

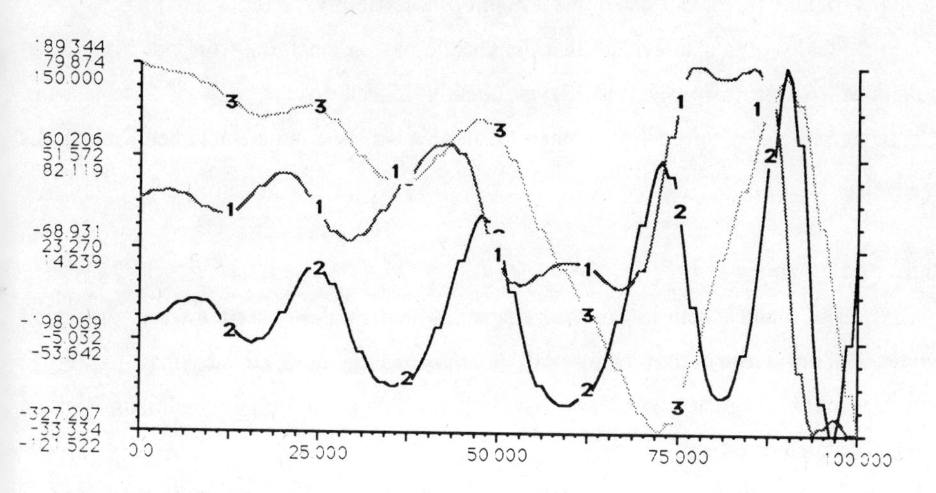

Figure 6.16 Chaotic Motions with Unfeasible Values for Population
y-axis = 1: employment rate; 2: immigration rate; 3: population (in descending order from the top)
x-axis = time

6.4.3 Concluding remarks

These simulation experiments show principally that our model of urban dynamics - in particular of urban decline - is a structurally unstable system, very sensitive to small changes in the parameters and in the initial conditions, and hence with the property of displaying oscillating behaviour. Clearly, the case of urban decline does not hold for all cities, but is still a rather common phenomenon in a certain phase of urban evolution: "empirical evidence is compelling in showing that the past 30 years or so many city centres lost population, whereas many suburbs gained population" (Dendrinos, 1980, p. 98). However, although urban decline can show oscillating results, our simulation experiments show that chaos, verified from a mathematical point of view, is not such a frequent state in the spatial activities.

Secondly, in these examples we can observe the importance of the parameter values, in particular of δ_3. Most attention certainly has to be paid to the speed of change of δ_3 since for $\delta_3 > 0.035$ we get unfeasible values of our variables, as well as to the spatial variables which can determine a high δ_3 (accessibility, facilities, etc.).

Finally, it is also evident that the chaotic pattern emerging from our interrelated equations (6.11), (6.14) and (6.17) is directly related to the size of urban areas. Consequently, when considering a small urban area we get a more stable behaviour in its decline.

6.5 Epilogue

The main conclusion of this chapter is that **models based on the theory of chaos do not ensure that chaos will actually occur**; however, we may expect an emergence of irregular dynamic behaviour, depending on initial conditions and on critical parameter values.

In this context much emphasis has to be put on the analysis of the speed of change of the parameters since some critical parameter values can lead to a chaotic movement with unfeasible values for some economic variables. Also the predictive value of such chaotic models has to be questioned, since - despite their deterministic nature - they may lead to unexpected results.

A further point concerns the type of chaotic models which are realistic in an economic-geographic context. For example, the logistic equation of the May-type is very simple in nature but it does not contain feedback effects which are more representative of real world phenomena. In this respect our generalized Lorenz model, applied to urban decline, is richer in scope. However, owing to feedback effects that often stabilize a system, it is not so easy to find critical values of the parameters leading to a Hopf bifurcation in our model (and hence a 'mathematical' route to chaos) with feasible values for the variables concerned.

This points out a final important issue, viz., the question of the **overall stability of a system's model**, if one of the subsystems is governed by chaotic motion. An example will be given in Chapter 8, where the impact of the logistic growth of May-type will be studied in a broader system of the competing dynamics of an ecologically based model. In this context, it should once more be noticed that the models analyzed in this chapter are macro models.

The possibility of a chaotic movement at a micro level will also be discussed in the next chapter, where the interrelations between DCMs and chaos theory, as well as the impact of lag-effects in the May-type equations, will be analyzed. In conclusion, the above mentioned research questions on chaos theory seem extremely relevant in economics and geography where local and sometimes global perturbations are plausible. Therefore, from this perspective, chaos theory provides new departures for research on dynamics of spatial systems behaviour.

Annex 6A

Classification of Two-dimensional Critical Points

In the first stage of a dynamic analysis, information on the state of the system can in principle be obtained on the basis of their equilibrium points and hence their (in)stability. Therefore, in this annex an analysis related to the classification of equilibrium points for a non-linear system of two differential equations will be given.In subsequent chapters we will refer also to the classification and the definitions used here.

In an autonomous system of two differential equations:

$$\frac{dx}{dt} = F_1 (x, y)$$

$$\frac{dy}{dt} = F_2 (x, y) \qquad (6A.1)$$

the point E_o (x_o, y_o) where F_1 (x, y) = F_2 (x, y) = 0, is called a critical point, steady state, equilibrium point or stationary point.

An equilibrium is stable if $\lim_{t \to \infty} x(t) = x_o$ and $\lim_{t \to \infty} y(t) = y_o$. In a more elaborate sense and following Kaplan (1958, p. 416), we may state that an equilibrium point E_o (x_o, y_o) is stable if: "for each l>0,a m>0 can be found so that every solution passing through a point (x_1, y_1) within distance m of (x_o, y_o) at time t_1 remains within distance l of (x_o, y_o) for all $t > t_1$ and approaches (x_o, y_o) as $t \to +\infty$. If the preceding conditions are satisfied, except for the requirement that the solutions approach (x_o, y_o) as $t \to +\infty$, then the equilibrium point is termed neutrally stable. If the equilibrium point is neither stable nor neutrally stable, then it is called unstable".

The basis of the classification of the equilibrium points for system (6A.1) can be found among others in Kaplan (1958), Ku (1958) and Oertel (1984). The most typical cases of the real eigenvectors and corresponding eigenvalues are illustrated in Figure 6A.1, where the following symbols are introduced:

$$\text{Trace} = p = \partial F_1 / \partial x \,|_0 + \partial F_2 / \partial y \,|_0 \qquad (6A.2)$$

$$\text{Jacobian Determinant} = q = \partial F_1 / \partial x \,.\, \partial F_2 / \partial y \,\big|_0 - \partial F_1 / \partial y \,.\, \partial F_2 / \partial x \,\big|_0 \qquad (6A.3)$$

$$\text{Discriminant} = D = p^2 - 4q \qquad (6A.4)$$

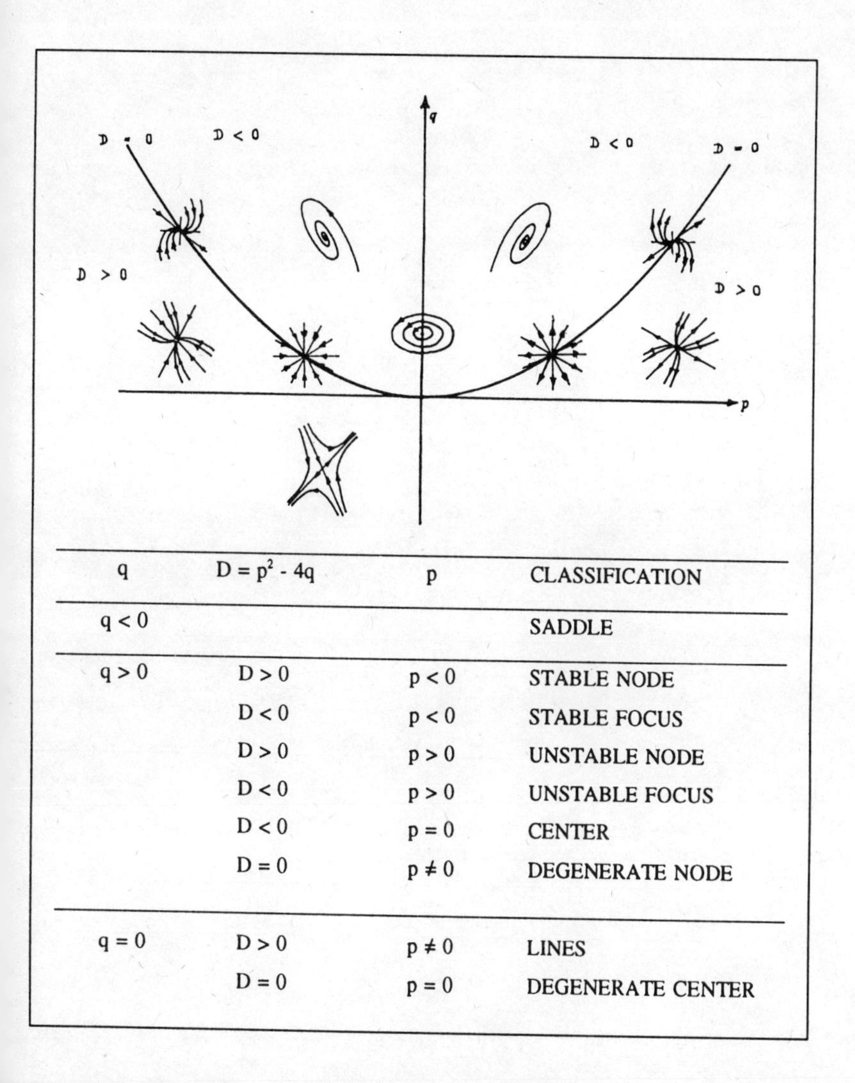

q	$D = p^2 - 4q$	p	CLASSIFICATION
$q < 0$			SADDLE
$q > 0$	$D > 0$	$p < 0$	STABLE NODE
	$D < 0$	$p < 0$	STABLE FOCUS
	$D > 0$	$p > 0$	UNSTABLE NODE
	$D < 0$	$p > 0$	UNSTABLE FOCUS
	$D < 0$	$p = 0$	CENTER
	$D = 0$	$p \neq 0$	DEGENERATE NODE
$q = 0$	$D > 0$	$p \neq 0$	LINES
	$D = 0$	$p = 0$	DEGENERATE CENTER

igure 6A.1 Classification of Two-dimensional Critical Points

Consequently, on the basis of information on the sign of trace p (defined in (6A.2)) and the sign of determinant q (defined in (6A.3)), it is easy to derive from Figure 6A.1 the specific type of instability related to our system (6A.1).

Annex 6B

Strange Attractors: A Brief Overview

In this annex a concise overview of the most relevant scientific contributions on strange attractors will be presented (see also Nijkamp and Reggiani, 1990c). Since the well-known example of May-attractor has been shown in Section 6.2.2 it will not be presented here. For further details on strange attractors we recommend some standard references: Collet and Eckmann (1980), Devaney (1986), Guckenheimer and Holmes (1983), Holden (1986), Iooss(1979), Lorenz (1989), Poston and Stewart (1986) and Schuster (1988).

6B.1 Hénon

Hénon (1976) has found a strange attractor in a two-dimensional quadratic mapping. In his contribution he clearly identifies a strange attractor by regarding it as a volume of a flow (in three-dimensional space) shrinking exponentially over time with a complex structure: "Inside the attractor, trajectories wander in an apparently erratic manner. Moreover, they are highly sensitive to initial conditions" (Hénon, 1976, p. 69).

Hénon starts from the following system of difference equations, describing a dynamic physical, chemical or biological system with two variables x and y:

$$
\begin{aligned}
x(t+1) &= y(t) + 1 - ax^2(t) \\
y(t+1) &= bx(t)
\end{aligned}
\qquad (6B.1)
$$

where a and b denote parameters. Given x(0), y(0) and by selecting particular values of a and b (a=1.4; b=0.3) he finds an attractor consisting of a number of more or less parallel 'curves' (see Figure 6B.1).

The Hénon attractor is a strange attractor because it is chaotic (i.e., with sensitive dependence on initial conditions); moreover, it has a **fractal** character (see Eckmann and Ruelle, 1985). Fractal is a term coined by Mandelbrot (1977), in order to illustrate that similar structures may repeat themselves at higher orders or dimensions. Essentially, a fractal set is a set having the property of being invariant at different scales (self-similarity and irregularity property) and having a non-integer, fractional dimension (for an application of fractal geometry to urban structure see, e.g., Batty and Longley, 1986, and Frankhauser, 1991). Therefore, the notion of fractals only refers to the geometry of attractors (see also Mandelbrot, 1977, and Peitgen and Richter, 1986).

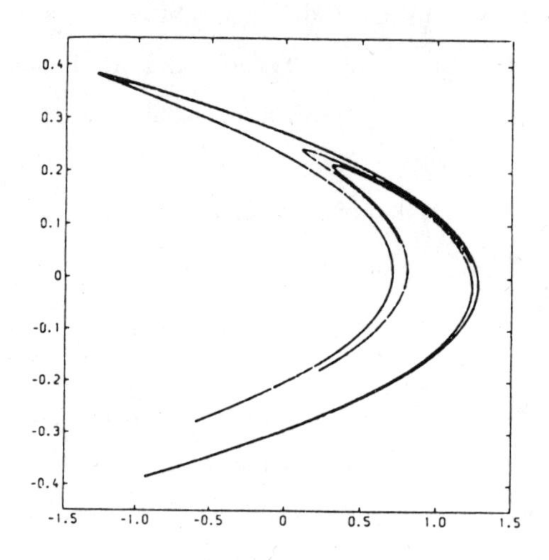

Figure 6B.1 The Hénon Attractor
Source: Hénon (1976, p. 73)

Since the notion of strange attractors usually refers to the dynamics of the attractors (and not just to their geometry), strange attractors need not have a fractal structure and attractors with a fractal structure need not be chaotic (see, e.g., Holden and Muhamad, 1986).

It should also be noted that - in contrast to a discrete time system - the minimal dimension (i.e., minimum number of equations) of a continuous time dynamical system which is able to generate chaotic time paths, is three. This result follows from the Poincaré-Bendixson theorem (see Section 6.2.2). The chaotic motion is also associated with the existence of homoclinic and heteroclinic orbits (see Sparrow 1982), i.e., orbits which either tend - both forward and backward - towards the same unstable stationary (saddle) point (i.e, homoclinic) or which tend - for both forward and backward movements - to different stationary points (i.e., heteroclinic).

It is interesting to illustrate next some further examples of strange attractors arising from continuous differential systems, rather than from discrete systems. This will be presented in the next three subsections.

6B.2 Gilpin

An interesting system of interacting species studied in an ecological context is the well-known Volterra-Lotka system of the following form:

$$[1/X_i(t)]\,\dot{X}_i(t) = g_i + \sum_j a_{ij} X_j(t) \qquad i=1,...,3; \quad j=1,...,3 \tag{6B.2}$$

Gilpin (1979) demonstrates that system (6B.2), which models the dynamics of a single predator X_1 and two prey species X_2, X_3 - (where $\dot{X}_i = dX_i/dt$ denotes the change over time of X_i, g_i mirrors the endogenous dynamics of each corresponding variable, and a_{ij} reflects the interaction between species), can give rise to chaotic trajectories (see Schaffer and Kot, 1986). Some illustrations of the Gilpin dynamic system (with the three variables X_1, X_2 and X_3 on the main axes) can be found in Figure 6B.2.

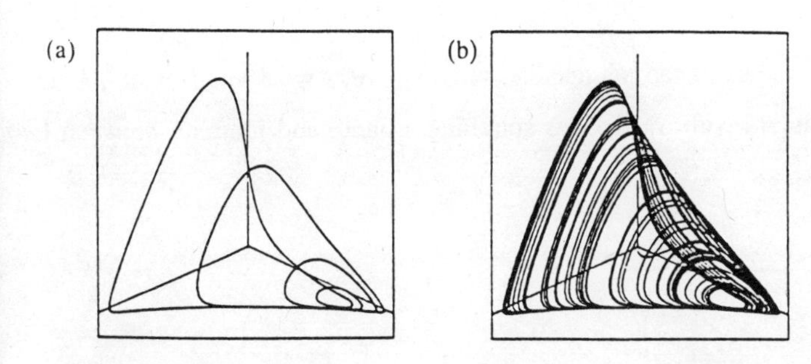

Figure 6B.2 The Chaotic Region for Gilpin's Equations:
(a) A4-Cycle; (b) A4-Cycle plus Complex Behaviour

Source: Holden (1986, p. 164)

In particular the chaotic region for Gilpin's equations also contains periodic orbits (see Figure 6B.2) which can be identified with the logistic ones.

6B.3 Lorenz

The Lorenz attractor is the first (and therefore the best-known) example of a strange attractor. Lorenz (1963) considers the following differential system describing a horizontal fluid layer heated from below and cooled from above, in an idealized two-dimensional rectangular flow region:

$$\dot{x} = \sigma(y-x)$$

$$\dot{y} = -xz + rx - y \qquad (6B.3)$$

$$\dot{z} = xy - bz$$

where in his original hydrodynamical system x represents the convective motion, y the horizontal temperature variation, and z the vertical temperature variation. The parameters σ, r and b are - because of their physical origins - usually assumed to be positive; in particular σ is the so-called Prandtl number (i.e., the ratio between thermal conductivity and kinematic viscosity of the fluid), r is the Rayleigh number (i.e., a dimensionless number proportional to the temperature difference between the two thermally conducting surfaces), and b is a number proportional to some physical proportions of the rectangle under consideration.

For particular values of the parameters, viz., $\sigma = 10$, $r = 28$ and $b = 8/3$, Lorenz finds complicated attractors with trajectories spiralling around, and jumping between two loops (see Figure 6B.3).

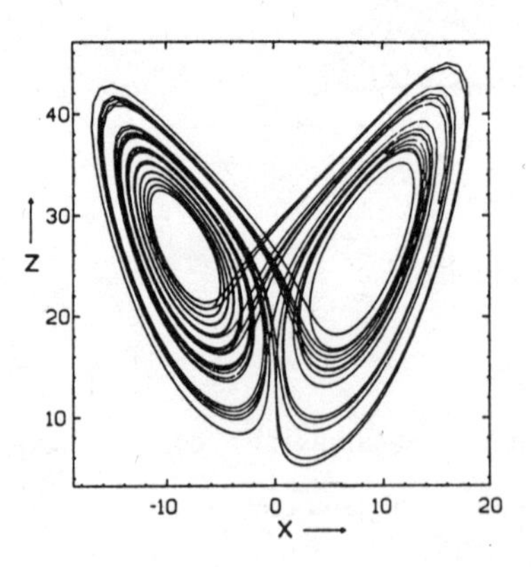

Figure 6B.3 Projection of the Lorenz Attractor for $r = 28$, $b = 8/3$, $\sigma = 10$
Source: Holden (1986, p. 126)

6B.4 Rössler

Rössler (1976) studied a simple three-dimensional system which models the flows around one of the loops of the Lorenz attractor:

$$
\begin{aligned}
\dot{x} &= -(y+z) \\
\dot{y} &= x + ay \\
\dot{z} &= b + z(x-c)
\end{aligned}
\tag{6B.4}
$$

where x, y, z refer to the three-dimensional motion of one loop emerging from (6B.4). In the classical form, the parameters a and b are treated as constants (a=b=0.2), while the parameter c is treated as a bifurcation parameter. Chaos develops at the accumulation point of the period-doubling sequence from a simple, period-one oscillation, just above c=4.2. However, also for other values of the parameters chaos may appear, leading to slightly different forms from the previous one (see Holden and Muhamad, 1986) (see Figure 6B.4). Rössler also developed from (6B.4) a four-dimensional system exhibiting a strange attractor with more complex trajectories (the so-called hyperchaos).

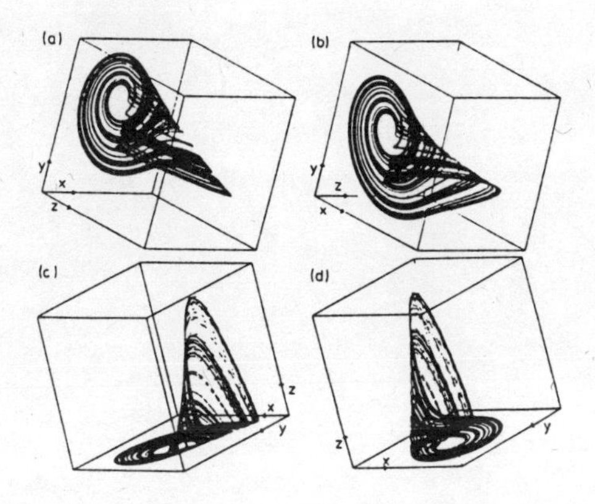

Figure 6B.4 The Rössler Attractor
Legend: Three-dimensional pictures with a = 0.343, b = 1.82 and c = 9.75; the axes form a cube which has been rotated around the y-axis (see (a) and (b)) and z-axis (see (c) and (d))
Source: Holden (1986, p. 24)

Finally, it should be noted that so far many software programmes exist, accessible on a microcomputer, which allow to examine the behaviour and the graphical illustrations of the main categories of dynamic systems. Two interesting examples worth mentioning here are DYNAMICS designed by J.A. Yorke, and PHASER by Koçak (1986).

Annex 6C

Steady State Solutions for a Generalized Lorenz System

In this annex we will investigate the existence of fixed point solutions for our system of generalized Lorenz equations. The conditions for a chaotic behaviour will also be discussed.

It is well known that the Lorenz equations presented in Annex 6B.3 possess the obvious basic steady state solution $x = y = z = 0$. With this solution the onset of convection is given for $r = 1$.

When $r > 1$, the Lorenz equations (6B.3) possess two additional steady states, viz.,:

$$x = y = \pm \{ b(r-1)\}^{1/2}$$
$$z = r-1 \qquad (6C.1)$$

around which - for particular values of the parameters - two spirals emerge (see also Figure 6B.3). We will now explore in an analogous way our dynamic system:

$$\dot{x} = -\sigma_2 x + \sigma_1 y$$
$$\dot{y} = -\sigma_2 xz + \delta_3 x - \delta_1 y$$
$$\dot{z} = \beta_2 xy - \beta_1 z \qquad (6C.2)$$

It is clear that also here the obvious steady state solution $x = y = z = 0$ exists. The linear transition matrix of (6C.1), based on a Taylor series around a steady state (x_o, y_o, z_o), is

$$\begin{pmatrix} x \\ y \\ z \end{pmatrix} = \begin{pmatrix} -\sigma_2 & \sigma_1 & 0 \\ \delta_3-\delta_2 z_o & -\delta_1 & -\delta_2 x_o \\ \beta_2 y_o & \beta_2 x_o & -\beta_1 \end{pmatrix} \begin{pmatrix} x-x_o \\ y-y_o \\ z-z_o \end{pmatrix} \qquad (6C.3)$$

Consequently, the characteristic equation of (6C.3) for the solution x = y = z = 0 is:

$$[K + \beta_1]\,[K^2 + (\sigma_2 + \delta_1)\,K + \sigma_2\,\delta_1 - \sigma_1\,\delta_3] = 0 \tag{6C.4}$$

It can easily be seen that the equation has three real roots when $\sigma_1\delta_3>0$. This system is stable, if $\sigma_1\delta_3 < \sigma_2\delta_1$. When $\sigma_2\delta_1 = \sigma_1\delta_3$, we have a critical value after which (unstable) motion begins (for $\sigma_1\delta_3 > \sigma_2\delta_1$). Furthermore, when $\sigma_1\delta_3 > \sigma_2\delta_1$, system (6C.2) possesses two additional steady state solutions, viz.,:

$$x = \pm\left(\frac{\beta_1}{\beta_2}\,\frac{\sigma_1\delta_3-\sigma_2\delta_1}{\delta_2\,\sigma_2}\right)^{1/2} = \frac{\sigma_1}{\sigma_2}\,y$$

$$z = \frac{\sigma_1\delta_3-\sigma_2\delta_1}{\sigma_1\,\delta_2} \tag{6C.5}$$

$$y = \frac{\sigma_2}{\sigma_1}\,x$$

The previous solutions may be tested on their stability conditions by means of simulation experiments (see also Section 6.4.2).

Next we show that the behaviour of the solution from (6C.5) is the same as in the original Lorenz model. In that case, a linearization of the positive vector field at the fixed point:

$$x = +\left(\frac{\beta_1}{\beta_2}\,\frac{\sigma_1\delta_3-\sigma_2\delta_1}{\delta_2\sigma_2}\right)^{1/2} = \frac{\sigma_1}{\sigma_2}\,y \qquad z = \frac{\sigma_1\delta_3-\sigma_2\delta_1}{\sigma_1\delta_2}, \tag{6C.6}$$

is:

$$M = \begin{pmatrix} -\sigma_2 & \sigma_1 & 0 \\ \frac{\sigma_2}{\sigma_1}\delta_1 & -\delta_1 & -\delta_2\left(\frac{\beta_1}{\beta_2}\frac{\sigma_1\delta_3-\sigma_2\delta_1}{\delta_2\sigma_2}\right)^{1/2} \\ \beta_2\frac{\sigma_2}{\sigma_1}\left(\frac{\beta_1}{\beta_2}\frac{\sigma_1\delta_3-\sigma_2\delta_1}{\delta_2\sigma_2}\right)^{1/2} & \beta_2\left(\frac{\beta_1}{\beta_2}\frac{\sigma_1\delta_3-\sigma_2\delta_1}{\delta_2\sigma_2}\right)^{1/2} & -\beta_1 \end{pmatrix} \tag{6C.7}$$

The characteristic polynomial of the matrix (6C.7) is:

$$K^3 + (\sigma_2+\delta_1+\beta_1)K^2 + (\frac{\beta_1\sigma_2^2 + \beta_1\sigma_1\delta_3}{\sigma_2})\,K + 2\beta_1\,(\sigma_1\delta_3-\sigma_2\delta_1) = 0 \tag{6C.8}$$

which has one negative and two complex roots.

It is straightforward to see, by applying the Hopf bifurcation theorem (see Marsden and McCracken, 1976), that for $\sigma_2>\beta_1+\delta_1$, a Hopf bifurcation occurs at the following point:

$$\delta_3^* = \frac{\sigma_2^2\,(\sigma_2+\beta_1+3\delta_1)}{\sigma_1\,(\sigma_2-\beta_1-\delta_1)} \tag{6C.9}$$

Obviously, (6C.9) may be a necessary but not sufficient condition for reaching chaos, since it is likely that the system possesses a strange attractor after the third Hopf bifurcation (see Section 6.2.1).

CHAPTER 7

SPATIAL INTERACTION MODELS AND CHAOS THEORY

7.1 Prologue

In the previous chapter the relevance of chaos theory, with special emphasis on spatial systems, has been demonstrated. In this context various important issues still deserve further attention. Firstly, a critical issue - according to Sterman (1988) - is that "the significance of the results hinges in large measure on whether the chaotic regimes lie in the realistic region of parameter space or whether they are mathematical curiosities never observed in a real system" (p. 148). Secondly, there are evidently major questions concerning the validity of both model specifications (notably, are model specifications compatible with plausible economic-spatial hypotheses) and of testability of model results (notably, are model results qualitatively or quantitatively justifiable from possible nonlinear patterns in the underlying data set).

It should be noted that the validity of model specifications is often associated with the behavioural rules postulated in various models of human systems. In particular, we recall here the debate on the compatibility between rationality and chaos which has indicated that chaotic solutions can occur in the presence of impatient economic agents (see Boldrin and Montrucchio, 1986), or incomplete, imperfect competitive markets (see Boldrin and Woodford, 1988). As a consequence "chaos has an interesting implication for the rational expectations literature. If the economy happens to be in a chaotic regime, then, even if economic agents know perfectly how the economy functions, they are unable to predict its behavior, except probabilistically" (cf. Pohjola, 1981, p. 37).

This chapter draws attention to this issue by extending the dilemma of rationality/chaos of actions to any behaviour of people with rational intertemporal choices. In particular, it will be shown that dynamic macro behaviour, derived from the maximization of a stochastic micro utility function as formalized in logit models, can exhibit chaotic or complex movements. Since dynamic logit models describe the choice probabilities among discrete alternatives, the fields of applications can be manifold, such as migration, transportation, residential choice, industrial location analysis, etc. (see, for example, Fischer et al., 1990, Hauer et al., 1989, Heckman, 1981, and Nijkamp, 1987b). Since complex behaviour seems to emerge for particular values of the (marginal) utility function of an actor, the present chapter also aims to treat the issue of the relationships between random utility theory and chaos theory and consequently the predictive power of discrete choice models derived from the above mentioned random utility theory. Furthermore, it will also be shown in this chapter that the introduction of multiple lags in a dynamic logit model may produce complex behaviour (under specific conditions of the utility function). This behaviour, however, appear to be bound to an even smaller region of the parameter space, as the number of delays increases.

7.2 Chaos in Spatial Interaction Models

7.2.1 Introduction

In this section the link between chaos theory and dynamic discrete choice models (and hence dynamic SIMs) will be examined. In particular it will be shown that under certain conditions for the utility function a dynamic MNL model in its difference form can exhibit chaotic behaviour emerging from a prey-predator system; moreover a degenerated logit model will be shown to display a chaotic pattern of the May-type (described in Section 6.2.2). This sheds new light on dynamic MNL models (and consequently on the equivalent SIMs), since here we will show that a dynamic logit model can generate (macro) randomness without any external random inputs. Therefore the present contribution aims to offer a new interesting insight into chaos analysis related to economics and geography as well as to dynamic analysis related to DCMs. Dynamic DCMs still represent an unexplored and open field of research, despite many recent advances (see for a review, Timmermans and Borgers, 1989). In this respect the present study can also be viewed as a new contribution to this research area.

7.2.2 Chaotic elements in dynamic logit models: theory

In Chapter 5 (see Section 5.2) it has been demonstrated that a dynamic MNL model may emerge as a solution to an optimal control problem whose objective function is a cumulative entropy function.

For the sake of simplicity we rewrite the spatial interaction expression (5.11) in the form of a dynamic logit model as follows (where the symbol t has been omitted):

$$P_{ij} = \frac{W_j \exp(-\beta c_{ij})}{\sum_l W_l \exp(-\beta c_{il})} \tag{7.1}$$

where $W_j = B_j^* D_j$ $(j = 1,...,l,...,J)$ is the weight (the so-called balancing factor) discussed in (5.11). Equation (7.1) can be easily transformed (by supposing for the sake of simplicity $W_j = 1$ and by omitting the symbol of the origin i) as follows:

$$P_j = \exp(u_j) / \sum_l \exp(u_l) \tag{7.2}$$

where u_j represents the utility of choosing alternative j $(j = 1,..,l,..J)$ (for instance moving to a place j) at time t.

Equation (7.2) is clearly the standard form of an MNL model in its multiperiod generalization; this is still, however, a comparative static model. We shall therefore consider the evolution of the dynamic MNL model (7.2) by writing the rate of change of P_j with respect to time (i.e., dP_j/dt):

$$\frac{dP_j}{dt} = \dot{P}_j = \frac{d}{dt}\left\{\frac{\exp(u_j)}{\sum_l \exp(u_l)}\right\} \tag{7.3}$$

Expression (7.3) leads, after simple computational exercises (see Nijkamp and Reggiani, 1991a) to:

$$\dot{P}_j = \dot{u}_j P_j (1 - P_j) - P_j \sum_{l \neq j} \dot{u}_l P_l \tag{7.4}$$

where $\dot{u}_j = du_j/dt$ represents the time rate of change of u_j. The latter term at the right hand side of (7.4) represents essentially interaction effects. By imposing now the condition $\sum_j P_j = 1$ in equation (7.4) we can derive from (7.4) that the related dynamic condition $\sum_j \dot{P}_j = 0$ is automatically satisfied.

Now it is interesting to examine in more detail possible specific trajectories of $\dot{u}_j$. By assuming, for instance, that the utility of choosing alternative j increases linearly with time through a fixed parameter α_j, we would have:

$$\dot{u}_j = \text{const} = \alpha_j \tag{7.5}$$

Then the evolution of an MNL model becomes:

$$\dot{P}_j = \alpha_j P_j (1 - P_j) - P_j \sum_{l \neq j} \alpha_l P_l \tag{7.6}$$

The surprising final expression is therefore a system of the prey-predator type (which will be extensively discussed in Chapter 8) with limited prey (P_j); P_l can be interpreted as a predator whose influence will be the reduction of population P_j through parameter α_l. It is interesting to notice that the first term in (7.6) represents the continuous time logistic equation depicting logistic growth (based, inter alia, on the Verhulst equation), while the second term of (7.6) represents interaction effects which may clearly affect the dynamic trajectory of (7.6).

We shall now approximate equation (7.6) in discrete time by considering a unit time period (see also Wilson and Bennett, 1985):

$$P_{j,t+1} - P_{j,t} \cong \alpha_j P_{j,t} (1 - P_{j,t}) - P_{j,t} \sum_{l \neq j} \alpha_l P_{l,t} \tag{7.7}$$

where α_j is now approximated by $u_{j,t+1} - u_{j,t}$, so that we have the following final expression in discrete terms:

$$P_{j,t+1} = (\alpha_j + 1) P_{j,t} - \alpha_j P^2_{j,t} - P_{j,t} \sum_{l \neq j} \alpha_l P_{l,t} \tag{7.8}$$

It should be noted that - given the discrete nature of most data in spatial analysis (such as survey data in transport decisions, or residential and migration choice data) - it seems plausible to specify spatial dynamic relationships of type (7.7) in difference form. For example, many trip decisions are made in a discrete sense (often represented as categorical data), so that there is no other choice alternative. Furthermore, there may be a delay effect in the response behaviour (see also the subsequent Section 7.3) which has to be modelled in discrete time intervals as well. Clearly, also condition

$\sum_j P_{j,t} = 1$ should be inserted in equation (7.7), although by adding the condition $0 \leq P_{j,t} \leq 1$ system (7.7) may not by definition generate real solutions (see Nijkamp and Reggiani, 1990d).

A further interesting observation is based on the analytical form of equation (7.8). The first two terms at the right hand side of equation (7.8) represent indeed a difference version of the standard discrete logistic growth model for a biological population X_t (with $X_t<1$), thoroughly discussed by May:

$$X_{t+1} = N X_t (1 - X_t) \tag{7.9}$$

where N is a parameter reflecting the growth rate ($0 < N < 4$) (See also Section 6.2.2).

By deleting for the time being the last term of (7.8) (i.e., the interaction term) and by assuming a constant utility increase (i.e., a_j), we find a degenerate case which is equivalent to (7.9). Now, if we simply put:

$$N_j = \alpha_j + 1 \tag{7.10}$$

we can rewrite equation (7.8) as follows:

$$P_{j,t+1} = N_j P_{j,t} (1 - \frac{N_j - 1}{N_j} P_{j,t}) \tag{7.11}$$

Thus equation (7.11) represents a **degenerated dynamic** MNL in discrete time. Now it could be interesting to compare the evolution of equations (7.9) and (7.11). It is evident that if we make the following transformation:

$$X_{j,t} = P_{j,t} (N_j - 1) / N_j \tag{7.12}$$

equation (7.11) can be written in the canonical form (7.9), i.e.,:

$$X_{j,t+1} = N_j X_{j,t} (1-X_{j,t}) \tag{7.13}$$

Here X_j assumes the meaning of the probability of choosing the first alternative in a binary choice situation; therefore we can derive the complementary probability as $(1-X_j)$. Consequently, the degenerated dynamic (discrete) logit model as specified in (7.11) belongs to the family of May models. This equivalence can also be verified also by comparing the related bifurcation diagrams (see Nijkamp and Reggiani, 1990b). **Thus our simple dynamic MNL model may in principle embody unstable or chaotic behaviour as soon as the model is specified in a difference equation form, at least in its degenerated version.** It is interesting to see that this happens for the values $3 < N_j < 4$ (or $2 < \alpha_j < 3$, being $\alpha_j = N_j - 1$); in other words, only when the slope α_j of the utility function u_j with respect to time (related to the dynamic logit model (7.11)) is less than 2 we have stable solutions in choosing the alternative j. Consequently, given the compatibility between MNL models and SIMs, **high values of the marginal utility function can lead to unpredictable movements** in spatial patterns (specified in a binary form) which are hardly foreseeable by modellers and planners. Clearly, if we would use the complete model specified in (7.8) alternative trajectories may emerge, depending on the evolution of the interaction term. In particular, since a discrete system of the prey-predator type can lead to a complex behaviour with strange attractors (see Peitgen and Richter, 1986, and Section 6B.2)) we expect that also in (7.8) chaotic movements arise. The evaluation of the effect of the interaction terms in the formulation (7.8) will be carried out by means of simulation experiments, supported by a mathematical analysis as will be illustrated in the next subsection.

7.2.3 Simulation experiments for a dynamic logit model

In this section we will show some simulation experiments related to the evolution of the discrete dynamic MNL model, represented by the prey-predator system (7.8). For this purpose we will examine the case of a system of three differential equations (i.e., the problem of the dynamic choice among three alternatives). In particular we will consider the dynamic logit model (7.8) both in relative terms (i.e., by inserting the additivity condition $\sum_j P_j = 1$) and in absolute terms in other words by considering the equivalent SIM). This analysis will illustrate that different types of behaviour (including chaotic movements) can emerge, depending on critical values of the utility function and initial conditions.

7.2.3.1 Dynamic processes in logit models

We will present here some results that demonstrate the stability/ instability of the dynamic logit model (7.8) (in relative terms). The importance of parameter α_j and the initial conditions will also be evaluated.

(a) Stable behaviour

In this simulation the following parameter values are considered:

$$\alpha_1 = 1.2 \qquad \alpha_2 = 1.1 \qquad \alpha_3 = 1$$

with the initial conditions:

$$P_1 = P_2 = 0.333 \qquad P_3 = 0.334.$$

Figure 7.1 shows a clear stable pattern for the system at hand.

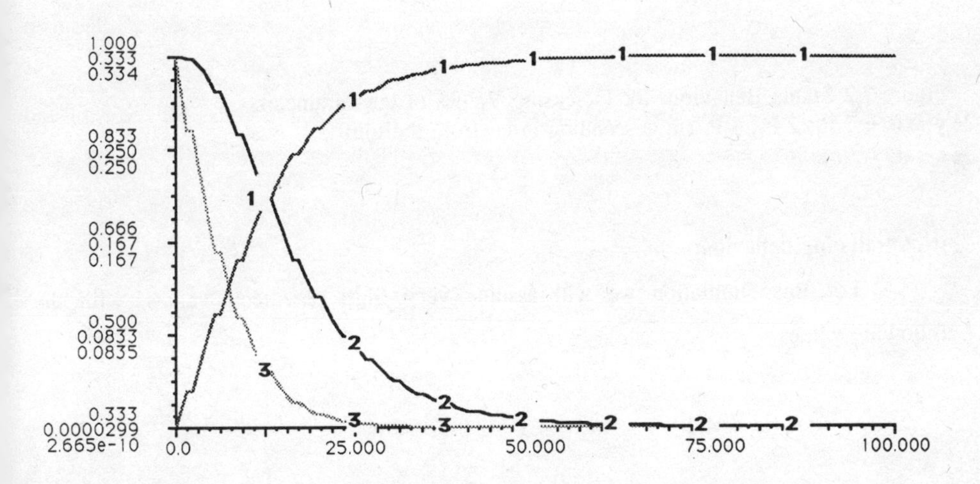

Figure 7.1 Stable Behaviour for a Dynamic Logit Model
y-axis = 1:P_1; 2:P_2; 3:P_3 (in descending order from the top)
x-axis = time

It is interesting to note the emergence of stable behaviour even for increasing values of α_j. If we consider, for example, the following values:

$$\alpha_1 = 2.9 \qquad \alpha_2 = 2.85 \qquad \alpha_3 = 2.8$$

with the same initial conditions ($P_1 = P_2 = 0.333$; $P_3 = 0.334$) we still obtain stable trajectories (see Figure 7.2). These two simulation experiments support our analytical results on stability discussed in Annex 7A.

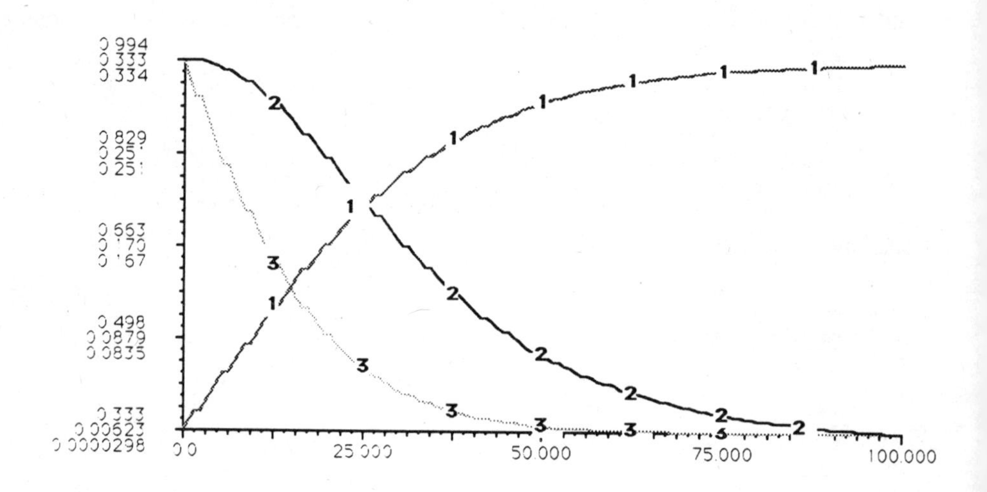

Figure 7.2 Stable Behaviour for Increasing Values of the Parameters
y-axis = 1:P_1; 2:P_2; 3:P_3 (in descending order from the top)
x-axis = time

(b) Oscillating behaviour

For this simulation we will assume **very high** parameter values, with the following values:

$$\alpha_1 = 4.2 \qquad \alpha_2 = 3.3 \qquad \alpha_3 = 6$$

with the same initial conditions as before:

$$P_1 = P_2 = 0.333$$
$$P_3 = 0.334$$

Figure 7.3 shows an oscillating behaviour for the whole time-period considered, as we also discussed in the equilibrium analysis carried out in Annex 7A.

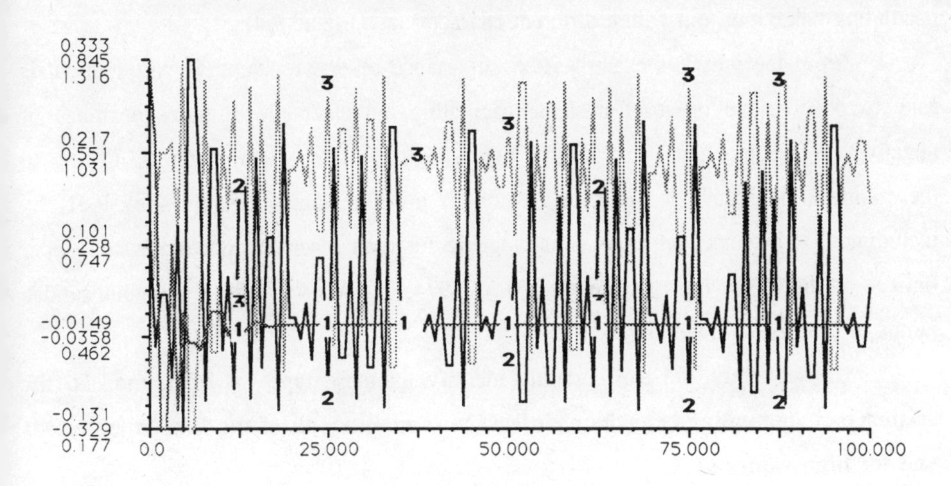

Figure 7.3 Oscillating Behaviour in Dynamic Logit Models
y-axis = 1:P_1; 2:P_2; 3:P_3 (in descending order from the top)
x-axis = time

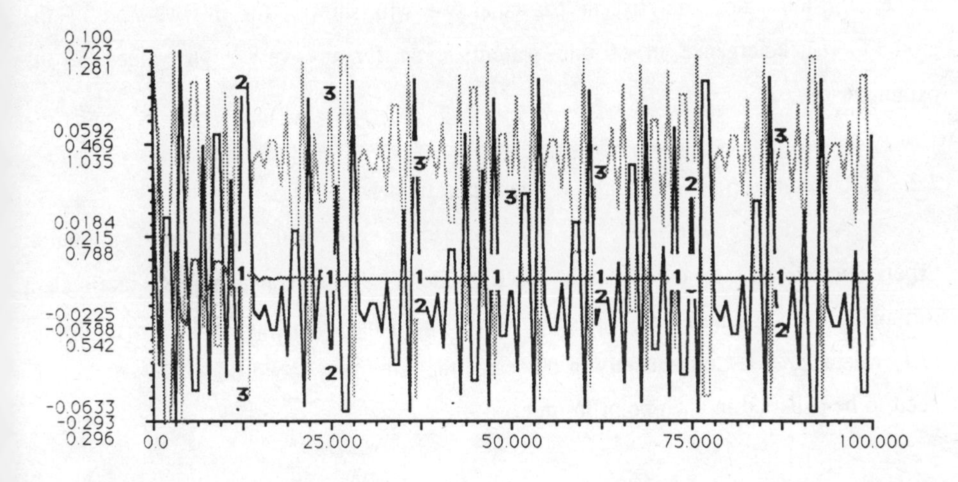

Figure 7.4 Oscillating Behaviour in a Dynamic Logit Model by Varying the Initial Conditions
y-axis = 1:P_1; 2:P_2; 3:P_3 (in descending order from the top)
x-axis = time

Next we can observe the sensitivity of the model with respect to the initial conditions by varying, in the last example, only the initial conditions of the variables (by assuming, for example, $P_1 = 0.1$; $P_2 = 0.45$; $P_3 = 0.45$); in this case we obtain again an oscillating behaviour, but with a different character (see Figure 7.4).

From the previous examples the importance of α_j is evident. In particular, it is easy to observe the onset of unstable oscillating movements with the appearance of negative (or greater than 1) values by increasing the value of parameter α_j. Owing to these unfeasible values we have to switch to positive values or values of $P_j<1$, by including, for instance, an exponential logistic function when P_j approaches the upper limit 1 or the lower limit 0 (see also May, 1974, and Wilson, 1981) to avoid sudden jumps in the system trajectory.

Consequently, we can certainly identify, in these types of MNL models, **the existence of fluctuating behaviour, at least in some intervals of the first time periods and for high values of α_j.**

A further step in the analysis of the evolution of the dynamic MNL model (7.8) is the investigation of its equivalent SIM (i.e., by considering the absolute value of the population P_j). In the following section we will illustrate some simulation experiments concerning the latter analysis; in particular we will show - for this version of the model - the emergence of chaotic patterns even for not very high values of the parameter α_j.

7.2.3.2 Dynamic processes in spatial interaction models

As already specified, in this sub-section we will carry out simulation experiments related to the dynamic MNL model (7.8), by considering its equivalent formulation in absolute terms; in other words the singly-constrained SIM of the form (7.1) where $B_j^* = 1$. Consequently in this case the additivity conditions $\sum_j P_j=1$ do not need to be satisfied in each point in time.

(a) Stable behaviour

In order to compare the dynamic trajectories in the two models (viz., the MNL model and the SIM), for this simulation we will assume the parameter values analyzed in the first example in 7.2.3.1.(a), i.e.,:

$$\alpha_1 = 1.2 \quad \alpha_2 = 1.1 \quad \alpha_3 = 1$$

with the initial conditions:

$$P_1 = P_2 = 0.333 \quad P_3 = 0.334$$

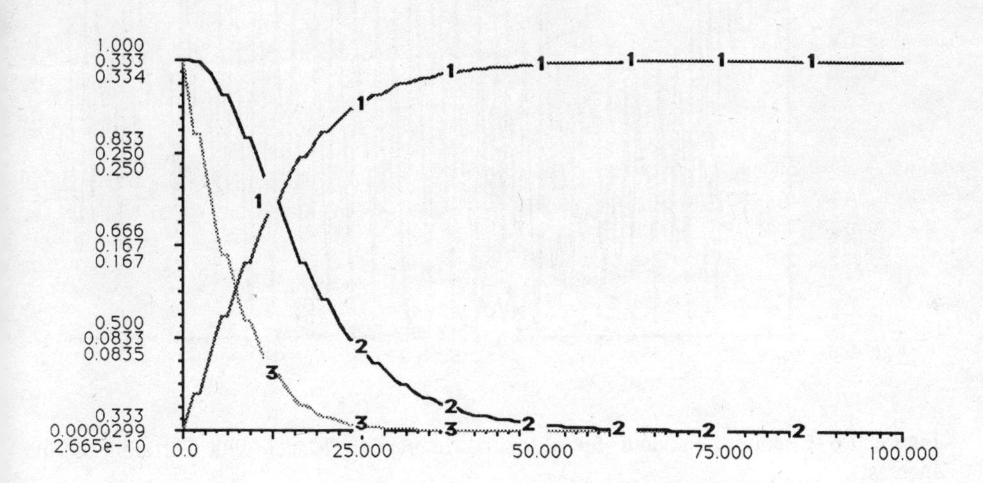

Figure 7.5 Stable Behaviour for a Spatial Interaction Model
y-axis = 1:P_1; 2:P_2; 3:P_3 (in descending order from the top)
x-axis = time

It is evident that under the previous conditions a stable behaviour emerges in the long run analogous to the one illustrated in Figure 7.1 (see Figure 7.5).

(b) Chaotic behaviour

In this simulation (where in each point of time the additivity condition $\sum_j P_j = 1$ is not explicitly imposed) we will assume the same parameter values related to the second example in 7.2.3.1.(a), i.e.:

$$\alpha_1 = 2.9 \quad \alpha_2 = 2.85 \quad \alpha_3 = 2.8$$

It is also assumed that the initial conditions satisfy nearly but not precisely the additivity condition $\sum_j P_j = 1$, i.e.:

$$P_1 = P_2 = P_3 = 0.333.$$

In this case the choice probabilities add up to 0.999, which is of course only slightly different from the ones used in the MNL case. The result can be found in Figure 7.6.

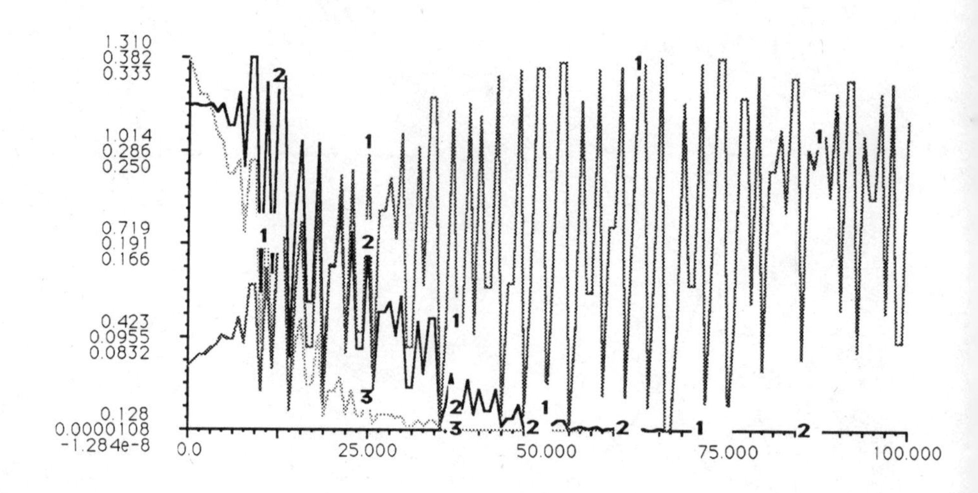

Figure 7.6 Chaotic Behaviour for a Spatial Interaction Model with Different Utility Increases
y-axis = 1:P_1; 2:P_2; 3:P_3 (in descending order from the top)
x-axis = time

Figure 7.6 shows the configuration of oscillating trajectories for the three populations of the SIM (7.8), whilst in the parallel case of an MNL model with the additivity condition $\sum_j P_j = 1$ we get stable behaviour (see Figure 7.2). This unstable behaviour is mainly the consequence of the values of the parameters that, in the case of a SIM, do not satisfy anymore the stability conditions (see Annex 7B). We can also notice from this pattern the influence of the utility increases α_j in the amplitude of the trajectories. In fact it is evident that populations P_3 and P_2 are the first variables reaching an equilibrium in the long run, since their related parameters α_3 and α_2 have the lowest values.

It is interesting to observe that when the parameter values are equal, the same type of chaotic pattern emerges for all three populations P_1, P_2, P_3. An example is given in Figure 7.7, where the following parameter values are considered:

$$\alpha_1 = \alpha_2 = \alpha_3 = 2.9$$

with the same initial conditions:

$$P_1 = P_2 = P_3 = 0.333.$$

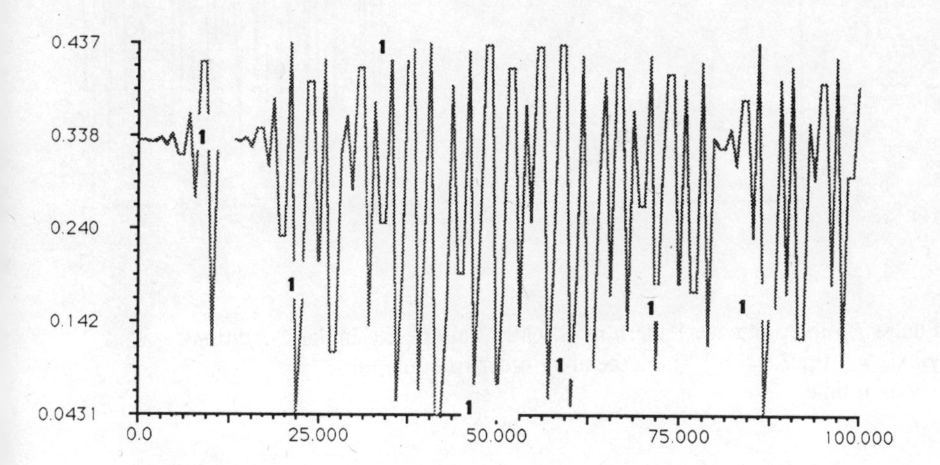

Figure 7.7 Chaotic Behaviour for a Spatial Interaction Model with Equal Utility Increases
y-axis = 1:P_1; 2:P_2; 3:P_3 (in descending order from the top)
x-axis = time

Furthermore, it is worthwhile to evaluate the impact of initial conditions on the above mentioned chaotic patterns. For this purpose we assume the initial conditions already used in the parallel case of an MNL model, i.e.:

$$P_1 = P_2 = 0.333 \qquad P_3 = 0.334$$

with the same parameter values previously examined in the MNL case (see the second example in 7.2.3.1(a)):

$$\alpha_1 = 2.9 \qquad \alpha_2 = 2.85 \qquad \alpha_3 = 2.8.$$

While in the parallel case of an MNL model we get a stable behaviour (see Figure 7.2), we obtain here an 'intermittency' pattern (see Figure 7.8).

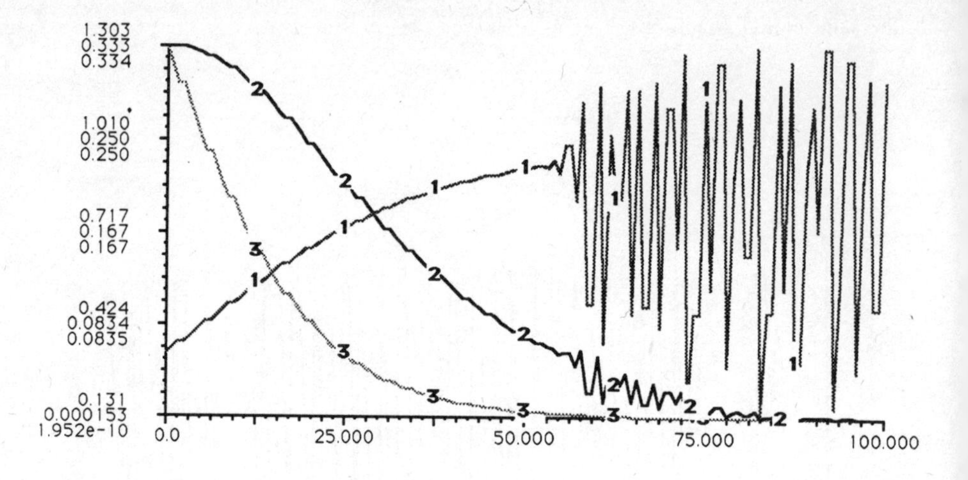

Figure 7.8 Intermittency Pattern by Slightly Varying the Initial Conditions
y-axis = 1:P_1; 2:P_2; 3:P_3 (in descending order from the top)
x-axis = time

In other words, after a period of stable behaviour a chaotic pattern arises. It is interesting to notice that in this example we have slightly changed the initial conditions compared to the first example of this subsection 7.2.3.2(b). Then, if we completely change the initial conditions, we can verify that again chaotic movements arise, characterized by new intermittency phenomena. Figure 7.9 shows a new chaotic evolution for the following initial values:

$$P_1 = 0.1 \qquad P_2 = P_3 = 0.45$$

with the same parameter values:

$$\alpha_1 = 2.9 \qquad \alpha_2 = 2.85 \qquad \alpha_3 = 2.8.$$

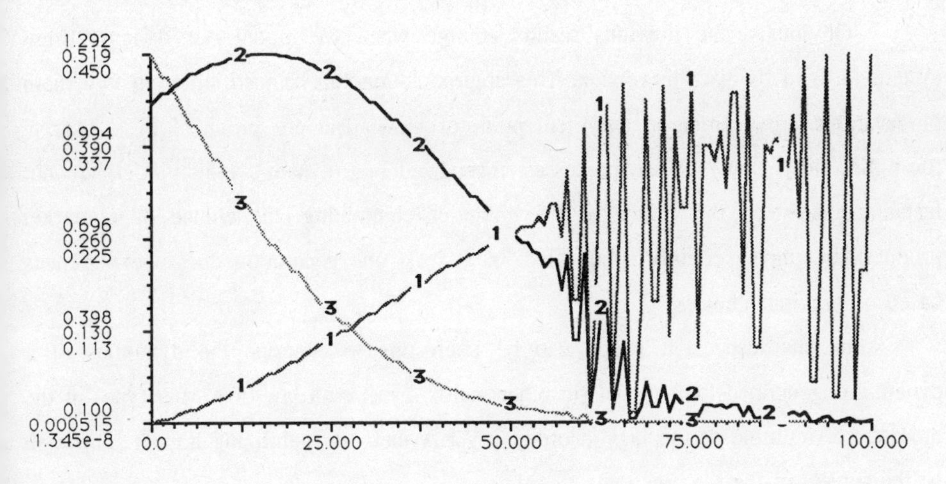

Figure 7.9 Intermittency Pattern by Completely Varying the Initial Conditions
y-axis = 1:P_1; 2:P_2; 3:P_3 (in descending order from the top)
x-axis = time

Once again we should note the importance of the utility increases α_j, since their high values seem to govern the instability of the trajectories.

7.2.4 Concluding remarks

In conclusion, chaotic behaviour seems to emerge in a dynamic discrete logit model, both in degenerated and non-degenerated cases. In the first case the dynamics follows a logistic chaotic pattern of the May-type, while in the second one the evolution reflects the behaviour of a prey-predator type. Furthermore, in this latter case - assuming the same values for the parameters and the initial conditions - the dynamic MNL in absolute terms (i.e., without the constraint $\sum_j P_j = 1$) shows more clearly an irregular pattern in comparison to the dynamic MNL in relative terms (i.e., by satisfying the additivity condition $\sum_j P_j = 1$). In both cases the growth parameters (i.e., the utility increases) play a fundamental role for the emergence of chaos. Consequently, much attention has to be paid to the **speed of change** of these **parameters**, since - as we showed - some critical parameters can lead to chaotic movements.

Obviously, the previous results emerge when we model our dynamic logit system (7.4) in its discrete version. This approximation can be justified from two main observations. First, from an empirical point of view, data are provided in a discrete form. Secondly, many choice processes, formalized in a dynamic logit model, are not instantaneous (e.g., the choice of the mode of commuting, the choice of a market product), but have certainly a lag-effect of at least one generation due to experiences based on previous choices.

In this respect, it could also be interesting to examine the dynamics of a dynamic degenerated logit model (in other words the growth law of a May type) in the presence of multiple period lags in order to verify their (de)stabilizing impact. This will be the subject of the next section.

7.3 Delay Effects in Dynamic (Binary) Logit Models

7.3.1 Introduction

In the previous section we have shown that a degenerated (binary) dynamic logit model belongs to the family of May equations (illustrated in Section 6.2.2). Consequently, we can apply May's results on stability also to binary dynamic logit models.

A further step along this research direction is the analysis of multiple delay effects in such a logit choice model (7.13) of a May type.

As noticed in the previous section, in many choice processes of actors, time lag effects are fundamental, since the decision process is often governed by a delayed response (i.e., the choice process is often not instantaneous), for instance, because of complicated learning and feedback effects. Moreover the study of delay effects in dynamic logit models allows one also to consider the interactions between people in space and time; these interacting choices cannot be incorporated in a static logit model owing to the well-known IID hypothesis (i.e., the random parts of the individual are Independent and Identically Distributed) (see Section 4.2.3.1).

In the context of a behavioural analysis interesting findings on the impact of the past have been obtained by Cugno and Montrucchio (1984a) in the context of adaptive expectations. Their results lead to the conclusion that chaos seems to vanish by increasing the weight of the past in rational expectations; however, even though the parameter related to delayed effects appear to be very high, there is always the possibility of unstable behaviour.

A further example underlying the relevance of the past history can be found in the area of urban analysis (see Johansson and Nijkamp, 1987), where the urban development follows a logistic discrete growth process characterized by a multi-episode history. In this context an event, i.e., the transition from one episode to another, is able to switch from stability to instability.

However, it should be noted that time delays in the growth dynamics of populations have been extensively studied, especially in mathematical ecology. We may refer here to the first work by Hutchinson (1948) and Maynard Smith (1974) who argue that if the duration of delays is longer than the natural period of the system (i.e. 1/N) divergent oscillations will result.

However, it is also increasingly realized that time delays are not necessarily destabilizing (see, e.g., Cushing and Saleem, 1982, and Hastings, 1983). Additionally, Saleem et al. (1987) confirmed this previous finding by arguing that increasing (decreasing) time delays are not necessarily destabilizing (stabilizing). It should be noted that these analyses have been mostly carried out by means of a two or three dimensional continuous system, such as the well-known prey-predator type of equations.

Starting from these previous results we will examine in this section the impact of multiple delays in the growth law of a May type, which can also be regarded as a formulation embedding a binary logit model in its difference form.

7.3.2 A logistic model with multiple delays

In order to take into account lag effects we will formalize here the following model, based on equation (7.13), by using the ecological specification introduced by Maynard Smith (1968):

$$X_{j,t+1} = N\, X_{j,t}\, (1-X_{j,t-n}) \tag{7.14}$$

where t-n indicates delay effects involving n time lags. In other words, we model the growth of the population share X_j choosing option j whose ability to grow in any given time span is governed by the population in the previous time span.

It is now clear that when n = 0, we get the first differential equation of a May type (see equation (7.13)), described in the previous section. Let us now consider the case of n = 1, i.e. the following equation:

$$X_{j,t+1} = N\, X_{j,t}\, (1-X_{j,t-1}) \tag{7.15}$$

A standard procedure in studying complex or chaotic behaviour is to draw a 'bifurcation diagram' in which the value of the variable is plotted against the parameter value. Consequently, if we then examine the bifurcation diagram emerging from (7.15) (see Figure 7.10), we observe a new shape, collecting cycles, and a cascade of bifurcations.

In particular the bifurcation diagram of Figure 7.10 confirms the equilibrium analysis of model (7.15), already investigated by several authors, such as Aronson et al. (1982), Lauwerier (1986), and Pounder and Rogers (1980). In particular these authors find unstable behaviour for the parameter range $2<N<2.27$, just showing for $N=2.27$ the existence of a strange attractor of Hénon type (see also Nijkamp and Reggiani, 1990d, 1992).

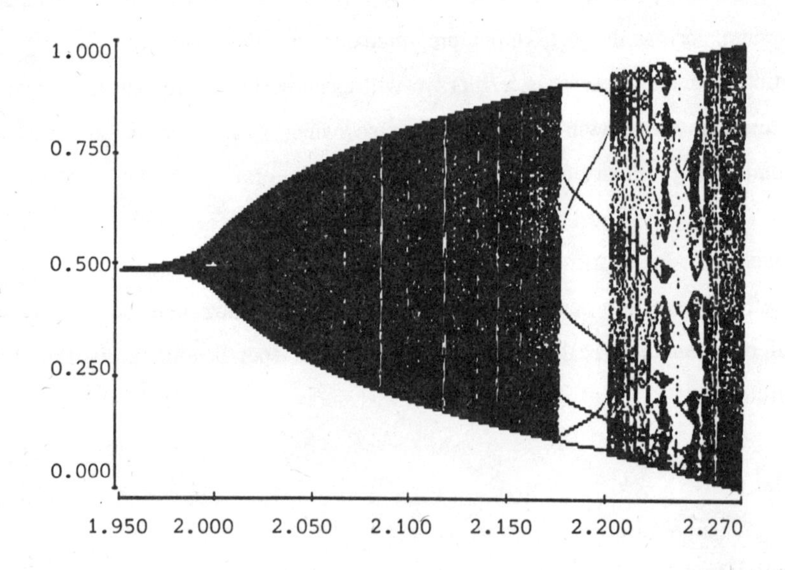

Figure 7.10 Bifurcation Diagram for the Logit Map with n=1
y-axis = X_j; x-axis = N

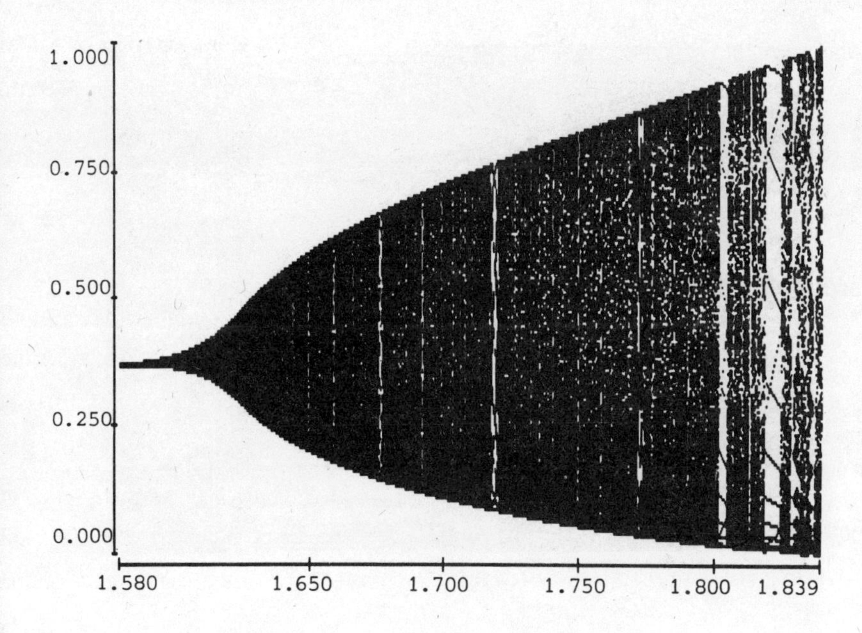

Figure 7.11 Bifurcation Diagram for the Logit Map with n=2
y-axis = X_j; x-axis = N

A further step is the investigation of more delays in equation (7.15). We will consider in particular the following values for n delays: n=2; n=5; n=10; n=21. The corresponding bifurcation diagrams are illustrated in Figures 7.11, 7.13, 7.17, 7.21, respectively, while Figure 7.12 shows an enlargement of Figure 7.11. Here, similar to the previous case, we assume that for the value of the growth parameter N=1.839 a strange attractor emerges, be it in three dimensions.

Then in Figures 7.14 and 7.16 we show some enlargements of the bifurcation diagram related to the value n=5, in particular near the area where the strange attractor likely occurs, i.e. at the end of the unstable (and feasible) region. In an analogous way we show in Figures 7.18 and 7.20 further enlargements of the dynamic behaviour near the strange attractor for n=10, while in Figures 7.22 and 7.23 we can see the 'nested' enlargement of the bifurcation diagram for n=21 (near the strange attractor). It is interesting to observe that, by increasing the number of delays, bifurcations result in a more compact way and show a regular structure (see, e.g., Figure 7.23 and the interesting blows up in Figures 7.15 and 7.19).

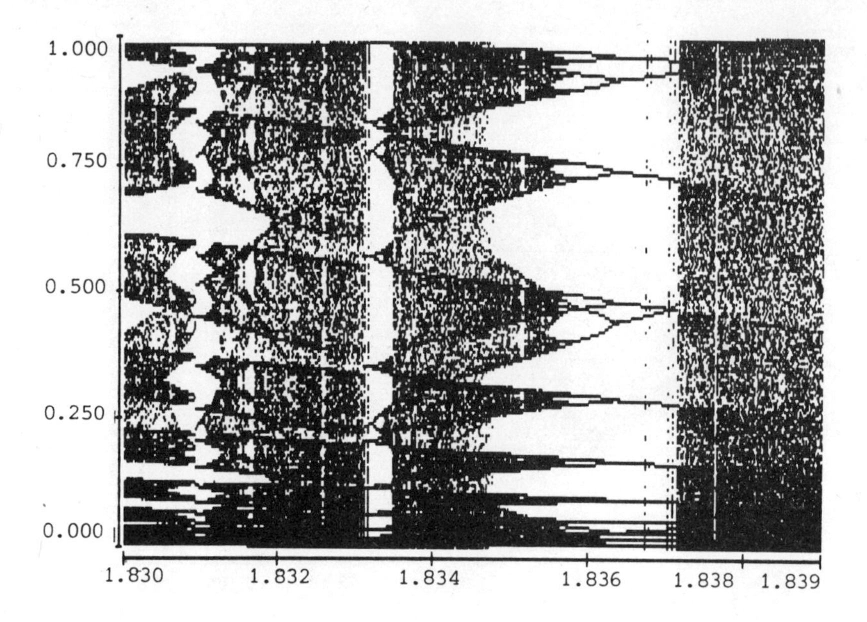

Figure 7.12 A Blow up of Figure 7.11

y-axis = X_j; x-axis = N

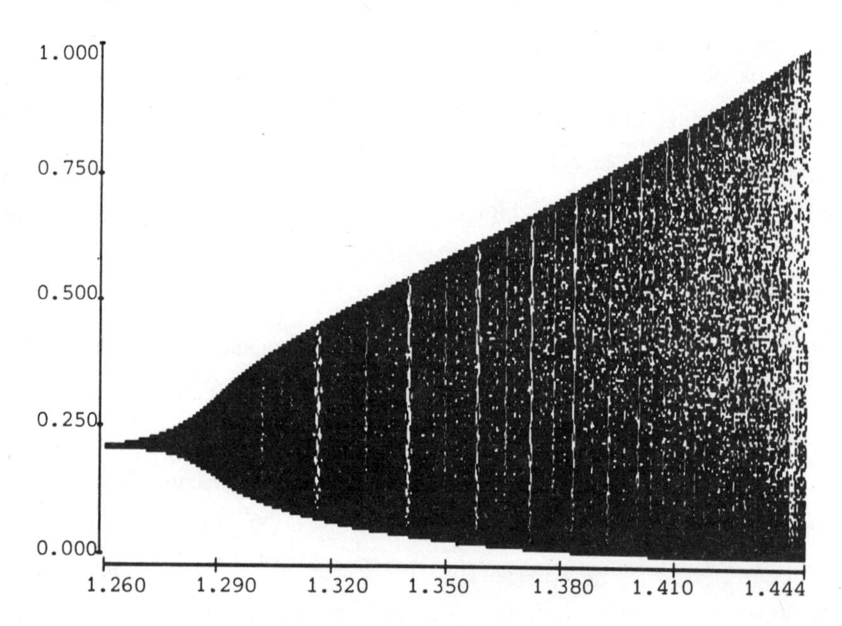

Figure 7.13 Bifurcation Diagram for the Logit Map with n=5

y-axis = X_j; x-axis = N

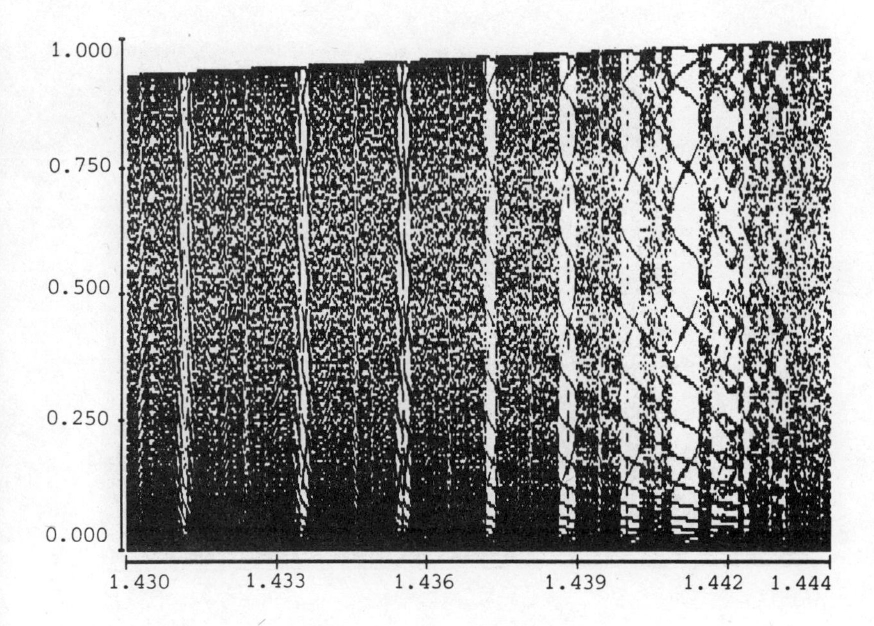

Figure 7.14 A Blow up of Figure 7.13

y-axis = X_j; x-axis = N

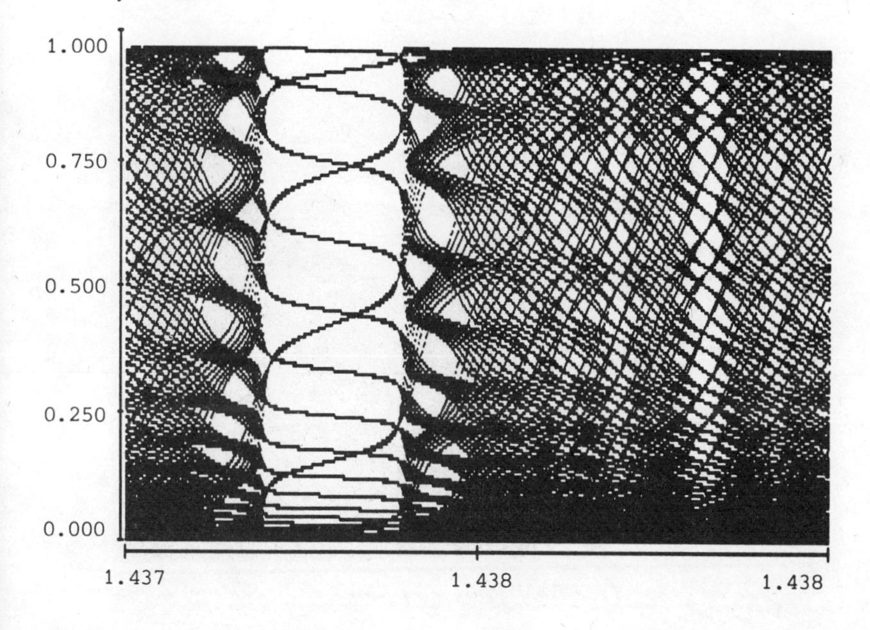

Figure 7.15 A Blow up of Figure 7.14

y-axis = X_j; x-axis = N

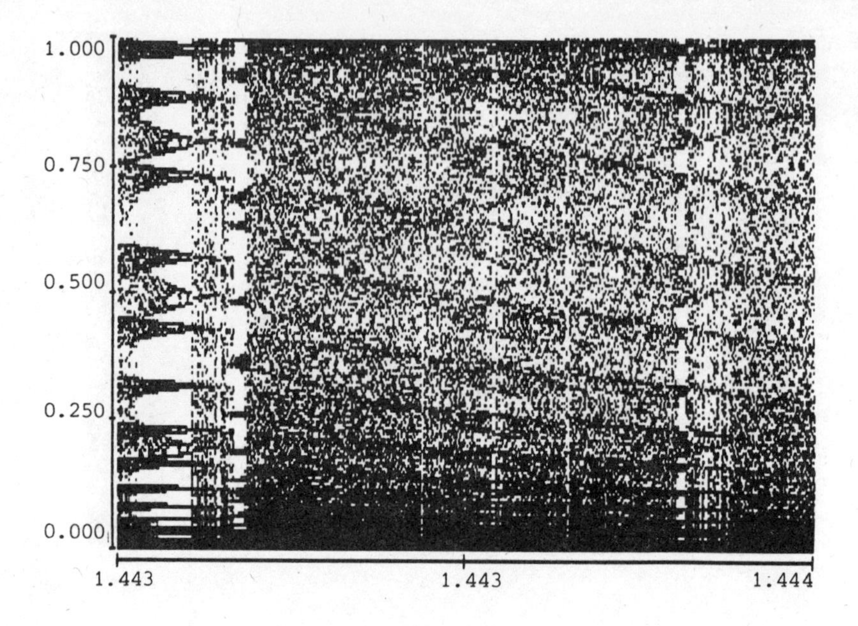

Figure 7.16 A Second Blow up of Figure 7.14 near the Strange Attractor
y-axis = X_j; x-axis = N

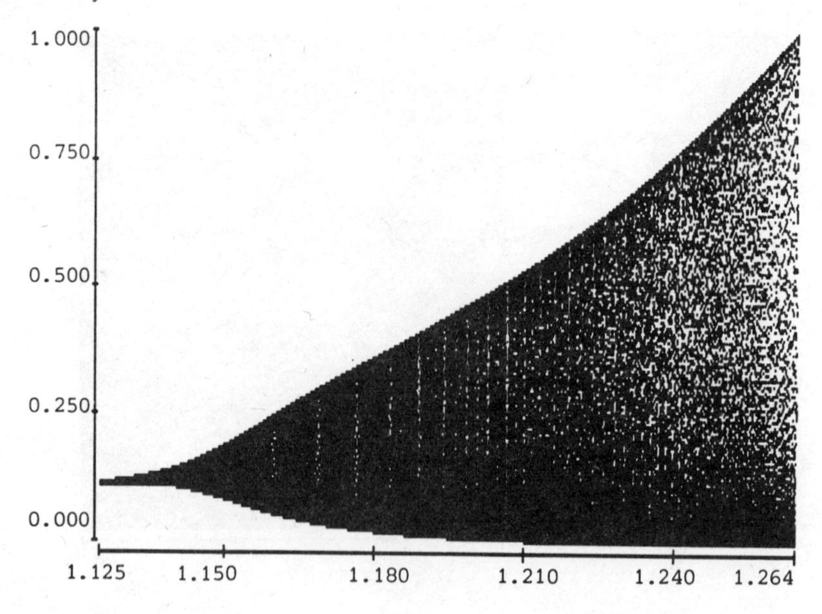

Figure 7.17 Bifurcation Diagram for the Logit Map with n=10
y-axis = X_j; x-axis = N.

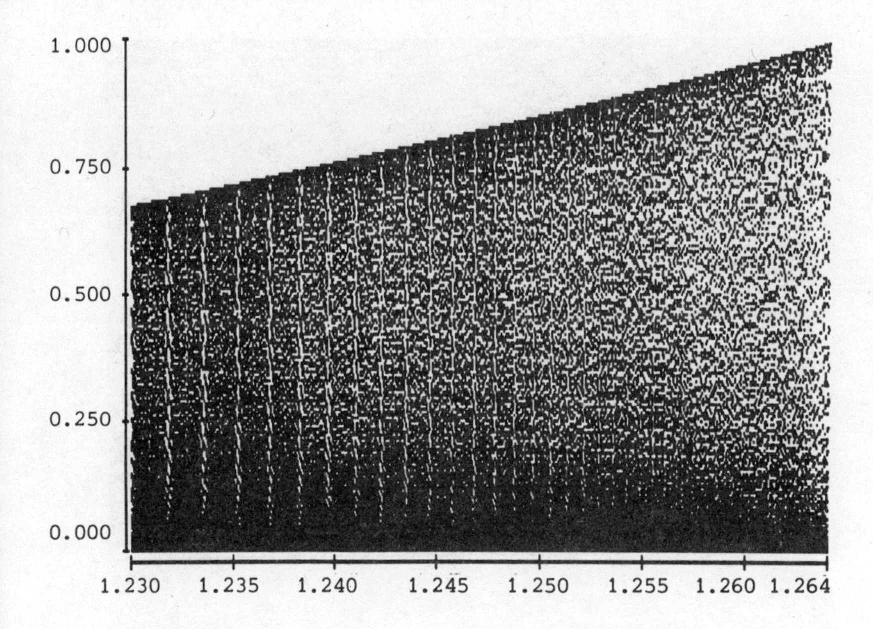

Figure 7.18 A Blow up of the Diagram Illustrated in Figure 7.17

y-axis = X_j; x-axis = N

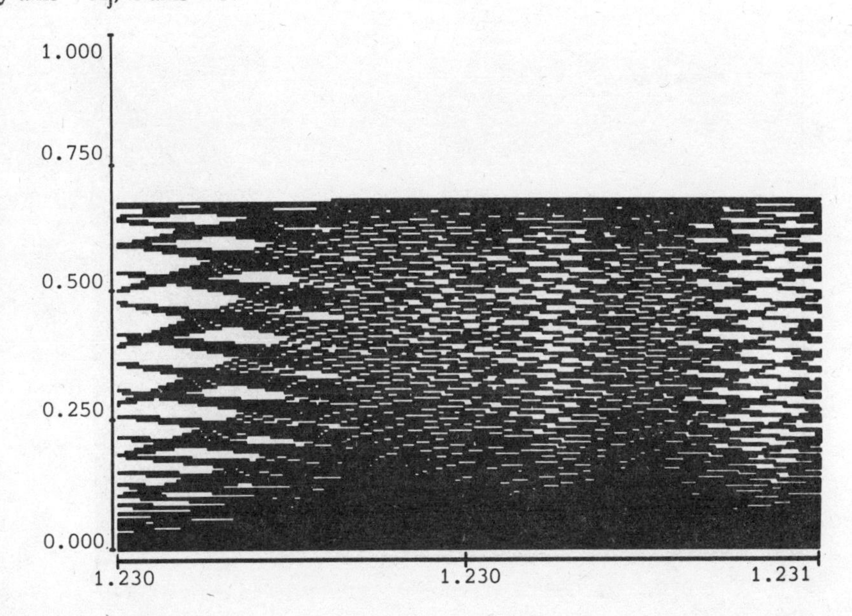

Figure 7.19 A Blow up of Figure 7.18

y-axis = X_j; x-axis = N

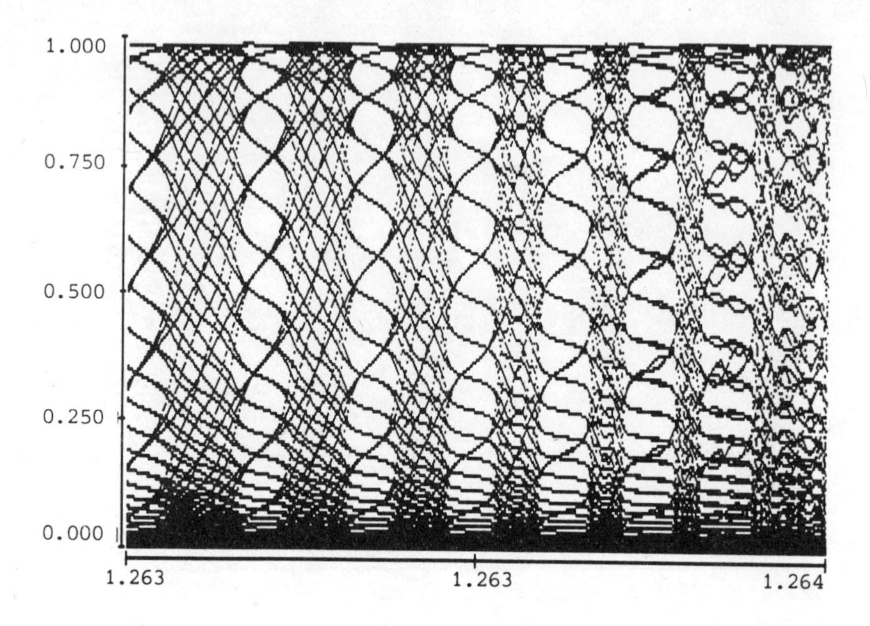

Figure 7.20 A Second Blow up of Figure 7.18 near the Strange Attractor
y-axis = X_j; x-axis = N

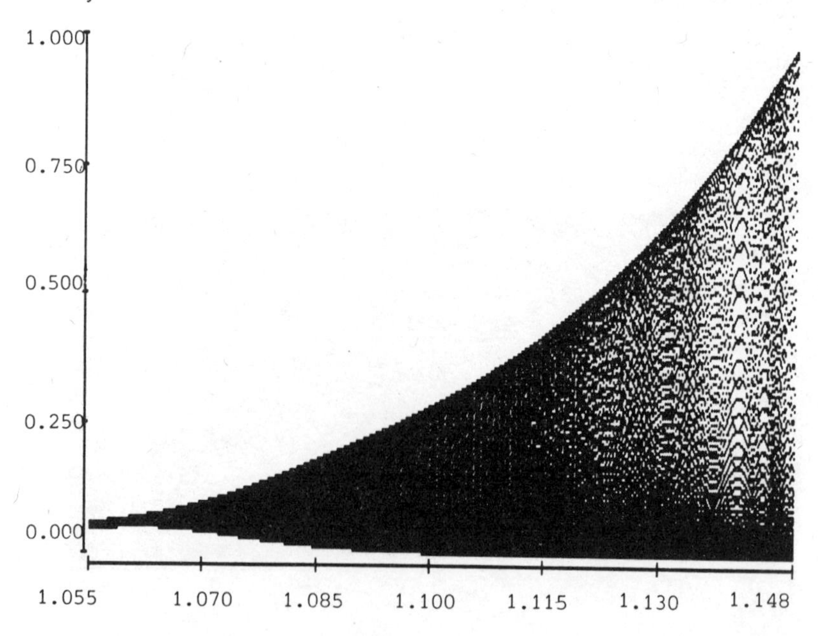

Figure 7.21 Bifurcation Diagram for the Logit Map with n=21
x-axis = X_j; x-axis = N

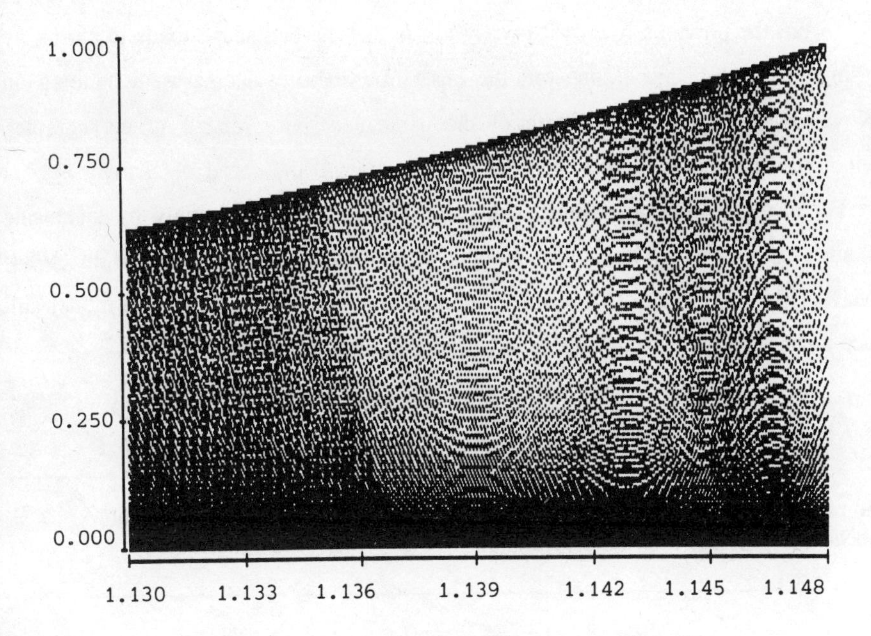

Figure 7.22 A Blow up of the Diagram Illustrated in Figure 7.21

y-axis = X_j; x-axis = N

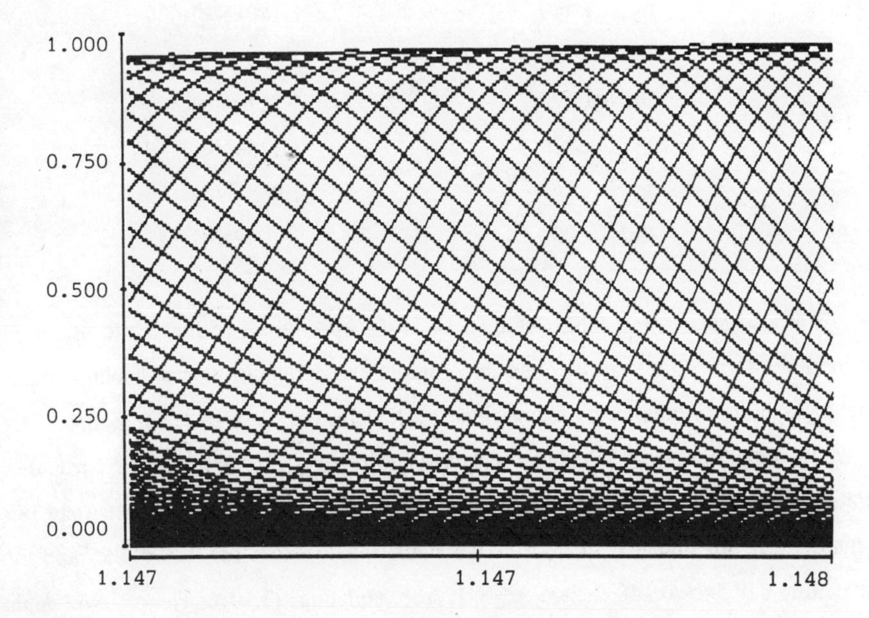

Figure 7.23 A Blow up of Figure 7.22 near the Strange Attractor

y-axis = X_j; x-axis = N

From the previous figures it is easy to see that the bell shape tends to shrink by increasing the time delays. Moreover, the onset of instability decreases, depending on the N values, as well as the length of the parameter space related to the complex behaviour. It also appears that as the delays are growing, the speed of change of N is lower. For example, the upper limit of instability decreases less strongly by increasing the number n of delays. This confirms some previous statements by Cartwright (1984) and May (1976) who claim that several variations on the logistic difference model still produce a chaotic regime, although at an earlier point (see also Table 7.1).

Table 7.1 Stable and Unstable Areas in Logit Maps for Increasing Delay Effects

n = number of delays	stable or periodic behaviour	complex behaviour with chaos
0	$0<N<3$	$3<N<4$
1	$0<N<2$	$2<N<2.27$
2	$0<N<1.580$	$1.580<N<1.839$
3	$0<N<1.440$	$1.440<N<1.639$
5	$0<N<1.260$	$1.260<N<1.444$
7	$0<N<1.170$	$1.117<N<1.345$
10	$0<N<1.125$	$1.125<N<1.264$
20	$0<N<1.055$	$1.055<N<1.155$
21	$0<N<1.055$	$1.055<N<1.148$

It is also clear from Table 7.1 that, since the upper limit of the stable region is equal to the inferior limit of the unstable region, the range of the parameter space leading to stability decreases by increasing the n delays. It is also evident that after many delays (approximately 21) stable and unstable regions are very thin. In particular, by relating the results emerging from Table 7.1 to the marginal utility function α of our logit map (7.13), we find that for $n = 21$ the stable region exists for $0<\alpha<0.055$ and the unstable for $0.055<\alpha<0.148$ (since $\alpha=N-1$; see equation (7.10)). Hence two main observations follow.

Firstly, for a large influence of the past very small changes in the utility function related to a choice process can lead to a switch from stable to unstable behaviour and vice versa. This result reinforces the conclusion reached by Hastings (1983) and Saleem et al. (1987) in an ecological model of a prey-predator type, as well as by Cugno and Montrucchio (1984a) in the context of adaptive behaviour, as already pointed out in Section 7.3.1. A link can also be found with the Johansson and Nijkamp model (1987) on population dynamics, where a new event (i.e., a small change) can transform the growth process from stability to instability, if there is a regulatory effect due to a large influence of the past.

Secondly, we also find that when the influence of the past is high, the conditions for the existence of a growth process of a logit type (depending on α) are very restrictive. For example, for n=21 the interval value necessary for feasibility is $0<\alpha<0.055$. This also means that, since α represents the variation of the utility function, by increasing the influence of the past the utility function tends to approach asymptomatically a constant value K, which may represent the maximum utility level due to the impact of the past. After having reached this value K, the past would then have no - or hardly any - more impact on the utility function.

7.3.3 Concluding remarks

In the previous experiments we have shown the impact of time specifications in growth models of a logit type (more precisely, in binary logit models) and hence in the related spatial models. In particular, we have shown that by increasing the number of time delays, such models tend to exhibit both regular and irregular behaviour (depending on the growth rates), although with more restrictive conditions on the growth rate for the feasibility of the model. Consequently, we have found the result - from a behavioural viewpoint - that after a certain number of generations the past does not influence any more the choice process.

From a 'structural' point of view we notice here - by increasing the n delays - the emergence of a 'regular' structure in the 'irregular' region of the bifurcation diagram that plots the probability of choice against the growth parameter. This is not a contradiction with the previous considerations where the increase of instability in a chaotic pattern gives rise to an 'ordered' structure.

7.4 Conclusions

In this chapter the possibility of chaotic phenomena in dynamic MNL models has been examined. Firstly it has been shown that both a dynamic MNL model and a dynamic delayed MNL model can be reduced to a **logistic form** (in the degenerate case of a binary choice) and hence may generate **chaotic behaviour** for particular values of the related utility functions as well as of the initial conditions. Moreover, the evolution of a dynamic logit model can be expressed in the general form of a prey-predator system, thus again displaying stable and complex behaviour for particular values of the utility function and initial conditions of the spatial interaction system.

Secondly, the influences of lag effects in dynamic binary logit models has been investigated. It resulted from our analysis that a dynamic binary MNL model with a delay of two generations can exhibit a strange attractor of the Hénon type with a fractal character. Further n delays lead to complex behaviour with likely the emergence of a strange attractor in n+1 dimensions. However, it seems that when n is high, the related 'highly chaotic' pattern shows an 'ordered' structure. This also seems to reinforce the general conclusion that after many delays the past has no more impact on the choice process.

In conclusion, the present chapter evokes an important research issue regarding the possibility of SIMs of describing not only equilibrium paths but also irregular behaviour; the latter situation is emerging in the presence of time lag effects. At this point it is certainly necessary to outline the broad scope of SIMs, in particular regarding their **potential** of also describing the 'casual' or 'chaotic' movements in real world situations.

A final remark concerns the question whether chaotic behaviour expressed by a logit map of a May type may be dampened by the dominance of a stable system or whether it will exert an explosive influence on it. A first step in this interesting research direction is offered in the next chapter, where the influence of a 'chaotic' evolution of a May type model in a broader system will be examined, in the particular context of ecologically-based models.

Annex 7A

Stability Solutions for a Dynamic Logit Model

In this annex we will analyze the stability properties of a dynamic logit model, in a discrete version as formulated in (7.7). For the sake of simplicity we will write system (7.7) as follows:

$$P_{j,t+1} = \alpha_j P_{j,t} (1-P_{j,t}) - P_{j,t} \sum_{l \neq j} \alpha_l P_{l,t} + P_{j,t} \tag{7A.1}$$

with the constraint:

$$\sum_j P_j = 1 \tag{7A.2}$$

Let us now consider on the basis of (7A.1) the case of a system of three differential equations in which we have inserted the constraint (7A.2):

$$\begin{aligned} P_{1,t+1} &= \alpha_1 P_{1,t} (P_{2,t} + P_{3,t}) - P_{1,t} (\alpha_2 P_{2,t} + \alpha_3 P_{3,t}) + P_{1,t} \\ P_{2,t+1} &= \alpha_2 P_{2,t} (P_{1,t} + P_{3,t}) - P_{2,t} (\alpha_1 P_{1,t} + \alpha_3 P_{3,t}) + P_{2,t} \\ P_{3,t+1} &= \alpha_3 P_{3,t} (P_{1,t} + P_{2,t}) - P_{3,t} (\alpha_1 P_{1,t} + \alpha_2 P_{2,t}) + P_{3,t} \end{aligned} \tag{7A.3}$$

It is now evident that the trivial F fixed points for which $P_{1,t+1} = P_{1,t}$; $P_{2,t+1} = P_{2,t}$ and $P_{3,t+1} = P_{3,t}$ are: F_0 (0,0,0) (unfeasible since $\sum_j P_j \neq 1$), F_1 (1,0,0), F_2 (0,1,0) and F_3 (0,0,1).

The local behaviour of the map (7A.3) at its feasible fixed points F_1, F_2 and F_3 is governed by its local linearization for which the Jacobian J taken at each fixed point $F(P_1^*, P_2^*, P_3^*)$ is the corresponding matrix. The Jacobian J from system (7A.3) (in which the constraint $\sum_j P_j = 1$ is inserted) is then as follows:

$$J = \begin{pmatrix} \alpha_1(P_2^*+P_3^*)-(\alpha_2 P_3^*+\alpha_3 P_3^*)+1 & \alpha_1 P_1^*-\alpha_2 P_1^* & \alpha_1 P_1^*-\alpha_3 P_1^* \\ \alpha_2 P_2^*-\alpha_1 P_2^* & \alpha_2(P_1^*+P_3^*)-(\alpha_1 P_1^*+\alpha_3 P_3^*)+1 & \alpha_2 P_2^*-\alpha_3 P_2^* \\ \alpha_3 P_3^*-\alpha_1 P_3^* & \alpha_3 P_3^*-\alpha_2 P_3^* & \alpha_3(P_1^*+P_2^*)-(\alpha_1 P_1^*+\alpha_2 P_2^*)+1 \end{pmatrix} \tag{7A.4}$$

Consequently, we get:

$$J[F_1\ (1,0,0)] = \begin{pmatrix} 1 & \alpha_1 - \alpha_2 & \alpha_1 - \alpha_3 \\ 0 & \alpha_2 - \alpha_1 + 1 & 0 \\ 0 & 0 & \alpha_3 - \alpha_1 + 1 \end{pmatrix} \tag{7A.5}$$

$$J[F_2\ (0,1,0)] = \begin{pmatrix} \alpha_1 - \alpha_2 + 1 & 0 & 0 \\ \alpha_2 - \alpha_1 & 1 & \alpha_2 - \alpha_3 \\ 0 & 0 & \alpha_3 - \alpha_2 + 1 \end{pmatrix} \tag{7A.6}$$

$$J[F_3\ (0,0,1)] = \begin{pmatrix} \alpha_1 - \alpha_3 + 1 & 0 & 0 \\ 0 & \alpha_2 - \alpha_3 + 1 & 0 \\ \alpha_3 - \alpha_1 & \alpha_3 - \alpha_2 & 1 \end{pmatrix} \tag{7A.7}$$

It is well known from literature that the eigenvalues λ_1, λ_2, λ_3 of each J (the so-called multipliers) should satisfy the following conditions for the stability of each fixed point:

$$|\lambda_1| < 1\,,\ |\lambda_2| < 1 \text{ and } |\lambda_3| < 1 \tag{7A.8}$$

Let us now consider, for example, the case of the fixed point $F_1(1,0,0)$. Its corresponding Jacobian (7A.5) shows that we are dealing with a triangular matrix so that the related multipliers are clearly the following:

$$\lambda_1 = 1\ ;\quad \lambda_2 = \alpha_2 - \alpha_1 + 1\ ;\quad \lambda_3 = \alpha_3 - \alpha_2 + 1 \tag{7A.9}$$

Consequently by combining (7A.9) with (7A.8) we get the following final conditions for stability:

$$\lambda_1 = 1$$

$$| \lambda_2 | < 1 \qquad (7A.10)$$

$$| \lambda_3 | < 1$$

From the second condition of (7A.10) we obtain the following values for α_2:

$$\alpha_2 > \alpha_1 - 1$$

$$\qquad (7A.11)$$

$$\alpha_2 < \alpha_1$$

or:

$$\alpha_2 < \alpha_1 - 1$$

$$\qquad (7A.12)$$

$$\alpha_2 > \alpha_1 - 2$$

In an analogous way we obtain the following stability conditions for α_3:

$$\alpha_3 > \alpha_1 - 1$$

$$\qquad (7A.13)$$

$$\alpha_3 < \alpha_1$$

or:

$$\alpha_3 < \alpha_1 - 1$$

$$\alpha_3 > \alpha_1 - 2 \qquad (7A.14)$$

If we now consider the values α_1, α_2, α_3 assumed in our simulation experiments related to sub-section 7.2.3.1, we arrive at the following conclusions. The parameter values in 7.2.3.1 (a), i.e., case 1) $\alpha_1 = 1.2$; $\alpha_2 = 1.1$; $\alpha_3 = 1$

case 2) $\alpha_1 = 2.9$; $\alpha_2 = 2.85$; $\alpha_3 = 2.8$

satisfy the stability conditions (7A.11) and (7A.13).

The parameter values in 7.2.3.1(b), i.e.:

case 3) $\alpha_1 = 4.1$; $\alpha_2 = 3.3$; $\alpha_3 = 6$

do not satisfy the stability conditions (7A.11)-(7A.14).

Simulation experiments carried out confirm the above formal analytical results on the stability of the fixed point $F_1(1,0,0)$.

Annex 7B

Stability Solutions for a Dynamic Spatial Interaction Model

In this annex we will analyze the stability properties of a dynamic spatial interaction model, in its discrete version by starting from the expression (7.7) without inserting explicitly the additivity conditions (7A.2). Consequently, in this case system (7A.3) leads to the following results:

$$
\begin{aligned}
P_{1,t+1} &= \alpha_1 P_{1,t} (1 - P_{1,t}) - P_{1,t} (\alpha_2 P_{2,t} + \alpha_3 P_{3,t}) + P_{1,t} \\
P_{2,t+1} &= \alpha_2 P_{2,t} (1 - P_{2,t}) - P_{2,t} (\alpha_1 P_{1,t} + \alpha_3 P_{3,t}) + P_{2,t} \\
P_{3,t+1} &= \alpha_3 P_{3,t} (1 - P_{3,t}) - P_{3,t} (\alpha_1 P_{1,t} + \alpha_2 P_{2,t}) + P_{3,t}
\end{aligned}
\qquad (7B.1)
$$

Also in this case it is evident that the trivial fixed points are: $F_0(0,0,0)$, $F_1(1,0,0)$, $F_2(0,1,0)$ and $F_3(0,0,1)$. The Jacobian J from system (7B.1) results then as follows:

$$
J=\begin{pmatrix}
\alpha_1-2\alpha_1P_1-(\alpha_2P_2+\alpha_3P_3)+1 & -\alpha_2P_1 & -\alpha_3P_1 \\
-\alpha_1P_2 & \alpha_2-2\alpha_2P_2-(\alpha_1P_1+\alpha_3P_3)+1 & -\alpha_3P_2 \\
-\alpha_1P_3 & -\alpha_2P_3 & \alpha_3-2\alpha_3P_3-(\alpha_1P_1+\alpha_2P_2)+1
\end{pmatrix}
\qquad (7B.2)
$$

Therefore, by considering - analogously to the previous Section 7A - the Jacobian related to the fixed point F_1 we get:

$$
J[F_1\ (1,0,0)] = \begin{pmatrix}
-\alpha_1+1 & -\alpha_2 & -\alpha_3 \\
0 & \alpha_2-\alpha_1+1 & 0 \\
0 & 0 & \alpha_3-\alpha_1+1
\end{pmatrix}
\qquad (7B.3)
$$

Jacobian (7B.3) is again a triangular matrix, and therefore the related multipliers are:

$$\lambda_1 = -\alpha_1 + 1\,; \quad \lambda_2 = \alpha_2 - \alpha_1 + 1\,; \quad \lambda_3 = \alpha_3 - \alpha_1 + 1 \tag{7B.4}$$

The stability conditions:

$$|\lambda_1| < 1\,;\ |\lambda_2| < 1 \text{ and } |\lambda_3| < 1 \tag{7B.5}$$

then require for $|\lambda_1| < 1$ the following conditions to be met:

$$\begin{aligned} &\alpha_1 < 1 \\ &\alpha_1 > 0 \end{aligned} \tag{7B.6}$$

or:

$$\begin{aligned} &\alpha_1 > 1 \\ &\alpha_1 < 2 \end{aligned} \tag{7B.7}$$

These results are in addition to the conditions on λ_2 and λ_3 that are exactly equal to (7A.11)-(7A.14) in the previous annex.

It is now clear that for the case of a SIM we get the following analytical results (by considering the same parameter values adopted for a logit model):

case 1) $\alpha_1 = 1.2$; $\alpha_2 = 1.1$; $\alpha_3 = 1$

These parameter values satisfy the stability conditions (7B.6) as well as the ones equivalent to (7A.11) and (7A.13).

case 2) $\alpha_1 = 2.3$; $\alpha_2 = 2.85$; $\alpha_3 = 2.8$

case 3) $\alpha_1 = 4.2$; $\alpha_2 = 3.3$; $\alpha_3 = 6$

These parameter values do not satisfy the conditions (7B.6) or (7B.7). Consequently, in these two cases the related system is unstable as has been demonstrated in our simulation experiments.

CHAPTER 8

SPATIAL INTERACTION ANALYSIS AND ECOLOGICALLY-BASED MODELS

8.1 Prologue

Ecologically-based models have recently become popular tools in dynamic systems analysis. In this chapter the possibility of analyzing SIMs in the context of (dynamic) ecological approaches will be examined from a social science viewpoint.

Mathematical ecology, with particular reference to Volterra-Lotka dynamics (see Volterra, 1926), has turned out to be a useful tool for studying the behaviour of regional or urban evolution. Volterra-Lotka dynamic equations provide rather simple models for the analysis of the interrelations between two (or more) species (or interlinked phenomena). These models can roughly be subdivided into three main groups: a) predation on species (prey-predator models); b) competition between species for limited food (competition for limited resources models); c) coexistence of species when different food resources are available (symbiosis models).

The ecological approach has recently also been used in urban and regional economics for analyzing competition between agents or activities in a limited space, e.g., the spatio-temporal diffusion of innovations (Sonis, 1983a and 1983b), the division of labour (Curry, 1981), the competition between two different types of production (Johansson and Nijkamp, 1987) and the competition among people for different locations (Dendrinos and Sonis, 1986). In this context various prey-predator models have also been developed, e.g., with reference to population and gross domestic product (Dendrinos, 1984b), to population and income (Dendrinos and Mullally, 1981) or to population and land price (Orishimo, 1987). This chapter is devoted to a further analysis of the relevance of such ecological models in the specific context of dynamic spatial interaction analysis.

The first part of this chapter will deal with the theoretical foundations of dynamic ecological models by discussing the stability solutions. Then a synergetic model of spatial interaction, based on the previous concepts, in the framework of an analysis of spatial flows of workers, will be developed. The assumption will be made that inflows and vacant workplaces can be interpreted as predator-prey phenomena, respectively. Next, an optimal control model will be formulated in order to study the stability conditions of the system concerned; the analysis will be carried out on the basis of a simple conventional utility function. It will be shown that the use of an optimal control model in this context may lead to a so-called 'borderline' stability with a possibility of bifurcation phenomena. Then the second part of the chapter will be devoted to the analysis of competition models with reference to regional systems. Although this type of models, by definition, does not display oscillating behaviour, it will be shown that a particular case of a 'chaotic' evolution of a May-type in one region will lead to an oscillating divergent behaviour in the entire system, if the intrinsic growth rates of the competing regions exceed a critical value (beyond which instability begins).

8.2 Prey-Predator Models: Introduction

The prey-predator model has originally been developed by Lotka and Volterra at the beginning of this century (see Lotka, 1925) as a tool for studying competitive behaviour between different species. In the two species case (with only one prey and one predator, represented by population size x_1 and x_2 respectively), a standard Volterra-Lotka model with limited prey can be written as follows (see for details Maynard Smith, 1974, and Volterra, 1926):

$$\dot{x}_1 = x_1 (b_1 - a_{11} x_1 - a_{12} x_2)$$

$$\dot{x}_2 = x_2 (-b_2 + a_{21} x_1) \tag{8.1}$$

The coefficients b_1 and b_2 are related to the endogenous dynamics of each corresponding variable, while the coefficients a_{11}, a_{12} and a_{21} reflect the interaction between species. Model (8.1) has two equilibrium solutions, a trivial one (viz., $x_1 = x_2 = 0$) and a more complicated one:

$$x_1^* = b_2 / a_{21}$$

$$x_2^* = b_1 a_{21} - (a_{11} / a_{12}) x_1^* \tag{8.2}$$

The Volterra-Lotka equations (8.1) cannot be solved in an analytical way due to their non-linearity, although solutions can be found for the linearized approximations (see Brouwer and Nijkamp, 1985).

However, if we restrict ourselves to the exploration of the optimal trajectories in a phase space diagram, we can plot the integral curves of model (8.1) for one given configuration of the coefficients (see Figure 8.1).

It is well known that for the case illustrated in Figure 8.1 (i.e., $b_1/a_{11} > b_2/a_{21}$), the equilibrium is stable (see, e.g., Hirsch and Smale, 1974, Wilson, 1981, and Wilson and Kirkby, 1980). But by varying the sign of the coefficient a_{11} (in particular by supposing a_{11} to be negative), we obtain a structurally unstable equilibrium (see Figure 8.2).

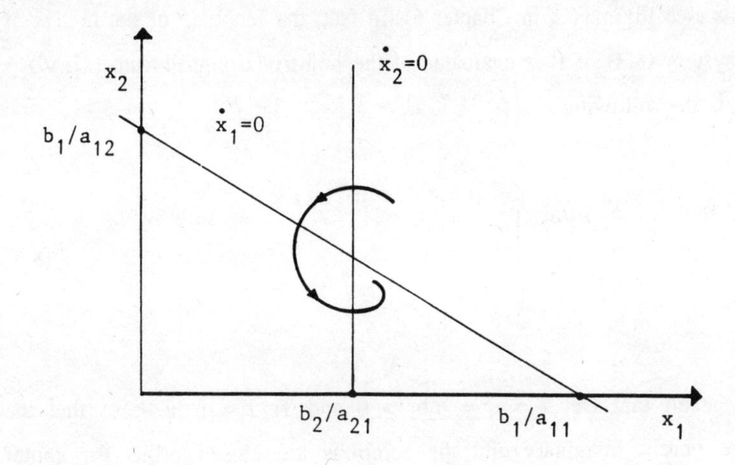

Figure 8.1 Stable Equilibrium for Prey-Predator Equations

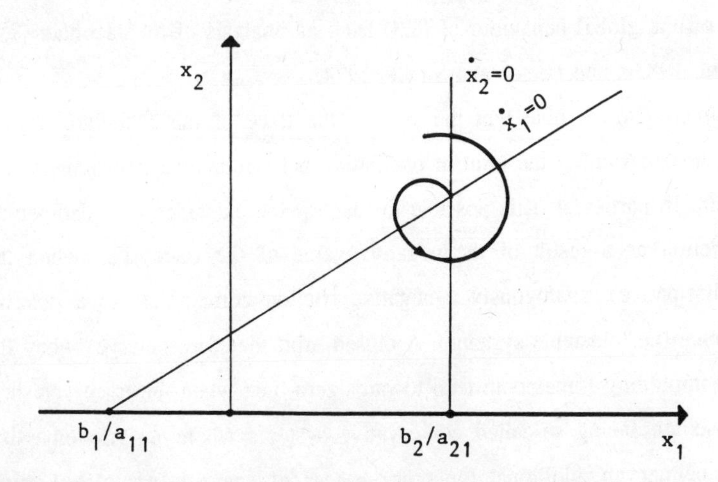

Figure 8.2 Unstable Equilibrium for Prey-Predator Equations

Therefore it is clear that the coefficient $a_{11} = 0$ represents a critical value at which a bifurcation may emerge - from stability ($a_{11} > 0$) to instability ($a_{11} < 0$). In particular it can be shown that when $a_{11} = 0$ the solutions of the system of differential equations:

$$\dot{x}_1 = x_1 (b_1 - a_{12} x_2)$$

$$\dot{x}_2 = x_2 (-b_2 + a_{21} x_1) \tag{8.3}$$

are closed orbits (see Figure 6.1 in Chapter 6). In fact, the Jacobian of the matrix of the linearized system (8.3), if it is evaluated at the non-trivial equilibrium $(x_1^*, y_1^*) = (b_2/a_{21}, b_1/a_{12})$, is the following:

$$J = \begin{pmatrix} 0 & -a_{12}b_2/a_{21} \\ a_{21}b_1/a_{12} & 0 \end{pmatrix} \tag{8.4}$$

leading to the result that Det $J = q = b_2 b_1 > 0$ and Tr $J = p = 0$, so that the eigenvalues are purely imaginary and the solutions are closed orbits (or center dynamics) with the center (as equilibrium point) at the origin (see also Annex 6A). System (8.3) is therefore neutrally stable implying that no unambiguous conclusion can be drawn on the global behaviour of (8.3) from an analysis of the Jacobian (8.4) (see also Haken, 1983a, and Hirsch and Smale, 1974).

It is interesting to note that the sign of the trace of the Jacobian plays a dominant role in determining the kind of oscillating behaviour of a two-dimensional dynamic system. In particular, it is possible to identify the existence of a dampening or positive friction as a result of the negative value of the trace (i.e., when the systems are dissipative); analogously a negative friction corresponds to a positive value of the trace (i.e., unstable systems). A closed orbit therefore emerges when the exploding and imploding forces both tend towards zero, i.e., when the trace vanishes. In this case we are facing so-called conservative systems where no friction exists (i.e., there is neither an additional input nor a loss of energy) (see also Lorenz, 1989).

It is clear that when we face conservative systems of type (8.3) we get temporal oscillations related to the two populations x_1 and x_2 as depicted in Figure 8.3.

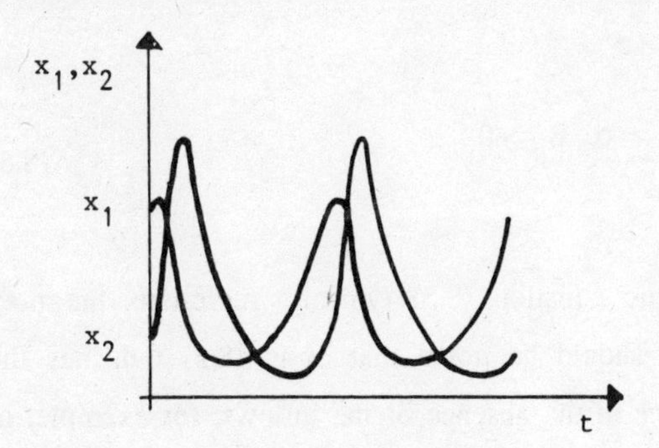

Figure 8.3 Time Variation of Two Populations (x_1, x_2), Corresponding to a Trajectory of Figure 6.1

Obviously in the case of competing species (i.e., when the rectangular terms are both negative) or symbiosis (i.e., when the rectangular terms are both positive), we do not obtain the periodic solutions as depicted in Figure 8.3 (or in Figure 6.1).

After this sketch of the theoretical background of ecological dynamic models we will now focus our attention on the possibility of developing dynamic SIMs on the basis of Volterra-Lotka equations. This will be the subject of the next section.

8.3 Synergetic Models of Spatial Interaction

The models described in the previous section (prey-predator, symbiosis, competing species) can be unified in the broad concept of so-called synergetic models referring in general to the cooperative interactions occurring between the various units of a system, in particular the impacts at a macroscopic scale (see, e.g., Haken, 1983a, 1983b, Sonis, 1983a, and Yamamura, 1987).

In the specific context of a **spatial** interaction system we will present now a dynamic spatial synergetic model on the basis of spatial cycles - in terms of growth and decline - between the attractiveness of labour markets (measured in terms of available new jobs or workplaces W_i) in a certain place i and related inflows into i from other areas j (i.e., T_{ji}), in the light of relevant **spatial** interactions between these areas.

The underlying hypothesis is that in a growing economy the number of new potential jobs (vacancies) on a given labour market i (denoted by $\dot{W}_i$) exhibits a growth pattern upon which the volume of inflows in i, T_{ji}, exerts a multiplicative negative impact:

$$\dot{W}_i = (\alpha_i - \sum_j \beta_{ji} T_{ji}) W_i \qquad \alpha_i, \beta_{ji} > 0 \tag{8.5}$$

It is clear that the dynamic equation (8.5) typically represents the prey phenomenon described in (8.3). It should be noted that α_i in (8.5) indicates the growth rate of the vacant workplaces in the absence of the inflows; for example, α_i might depend on new technological regimes. The coefficient β_{ji} represents the decline rate of W_i owing to various inflows T_{ji}. Obviously the parameter β_{ji} indicates also that not all inflows can be accommodated by the labour market because of lack of matching caused by differences in qualifications, etc.

The second assumption is that the growth of inflows T_{ji} is positively influenced by the number of vacant workplaces W_i, so that this equation is related to a predator phenomenon, as defined in (8.3):

$$\dot{T}_{ji} = (\sigma_{ji} W_i - \phi_i) T_{ji} \qquad \sigma_{ji}, \phi_i > 0 \tag{8.6}$$

In equation (8.6), ϕ_i represents the rate of decline in inflows T_{ji} in the absence of workplaces W_i, while σ_{ji} represents the interaction rate between the two populations (similar to β_{ji}).

It is interesting to observe that models of the form (8.5) and (8.6) bear some resemblance to the economic growth cycle model formalized by Goodwin (1967). In fact this author has analyzed a system of non-linear differential equations (of the (8.5) and (8.6) type), which describes both the motion of the employment rate and that of the workers' income share. Despite some criticism on the realism of this model approach, it still remains the basis for many theoretical contributions and empirical applications in the field of dynamic evolutions of economic systems (see, among others, Balducci et al., 1984, Funke, 1987, and Velupillai, 1983).

However, predator-prey models of the particular (8.1) type (i.e., with limited growth) occur much more frequently in the economic literature, mostly in fishery and other renewable resource models (see, e.g., Chauduri, 1987, and Ragozin and Brown, 1985). A reason for this popularity (see Hannesson, 1983) is that equations (8.1) are capable of producing a stable equilibrium, while the system described by equations (8.5) and (8.6) exhibits oscillating behaviour, as has been previously illustrated.

Obviously, models (8.5) and (8.6) assume that no dissipative force exists (e.g., the presence of congestion externality); otherwise we need a model of (8.1) type converging toward a steady state represented by a spiral sink. In other words, the models (8.5) and (8.6) represent the orbital dynamic equilibrium of the spatial system considered, as the effect of congestion disappears (see Figure 8.4). Consequently, such equilibrium behaviour requires an examination of the 'optimum path' within this framework. Such an optimum trajectory could be the result of maximizing some social benefits involved in reaching the steady state over a prespecified time horizon. This problem can be analyzed by evaluating an optimal control policy related to the oscillatory motion in the inflows and workplaces, as will be discussed in the next section.

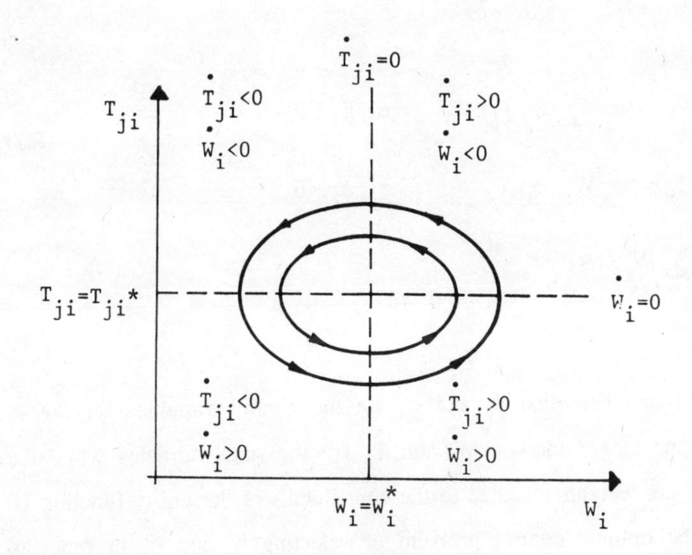

Figure 8.4 Spatial Cycles between Inflows and Workplaces

8.4 An Optimal Control Model for a Spatial Prey-Predator System

8.4.1 Introduction

In Chapter 5 the use of optimal control theory for dynamic SIMs has been advocated. In the present section this mathematical tool will be applied to an ecological approach to spatial dynamics. In this context it is interesting to recall the contribution by Dendrinos and Mullally (1983) to dynamic spatial systems analysis where also optimum control theory is used (on the basis of Goh et al.'s model (1974)), in the framework of urban dynamics based on ecological models of a prey-predator type reflecting the interaction between population and income, respectively. While Dendrinos employs an objective function incorporating per capita income and tax rate, we will adopt here a frequently used (concave) logarithmic utility function reflecting the interaction between inflows and workplaces (see for a general theoretical foundation Somermeijer and Bannink, 1973). Both the flows T_{ji} and the workplaces W_i are assumed to be arguments of such a collective utility function.

Therefore our optimal control model can be written as follows:

$$\max U = \int_0^{\tilde{T}} [\sum_i \sum_j \iota_i \ln (\beta_{ji} T_{ji}) + \sum_i \sum_j r_j \ln (\sigma_{ji} W_i)] \, dt$$

subject to

$$\begin{aligned} \dot{W}_i &= (\alpha_i - \sum_j \beta_{ji} T_{ji}) W_i \qquad && \alpha_i, \beta_{ji} > 0 \\ \dot{T}_{ji} &= (\sigma_{ji} W_i - \phi_i) T_{ji} \qquad && \sigma_{ji}, \phi_i > 0 \\ \sum_i W_i &= \bar{W} && \end{aligned} \tag{8.7}$$

We will assume here that β_{ji} and σ_{ji} are the control variables serving to enhance access to the labour market; W_i and T_{ji} are the state variables while the parameters τ_i and r_j are weights attached to the constituents of the utility function U. We will analyze the optimal control problem of selecting β_{ji} and σ_{ji} in order to maximize utility function U (i.e., a maximum interaction or matching potential between inflows and workplaces) during the finite time horizon $\tilde{T}$. We recall that β_{ji} is a compound parameter encompassing the effects of distance frictions, of heterogeneity on the labour market and of equilibrium mechanism (e.g., wages).

Problem (8.4) also takes into account the additivity condition $\bar{W}$ on a closed system, so that model (8.7) is essentially a pooling model for the labour market. This means that first the volume of workplaces is allocated and that next the resulting flows are determined.

Owing to the static constraints on W_i, problem (8.7) is a bounded optimal control model (see also Annex 5A), so that we have to use both the Hamiltonian and the Lagrangian function for deriving the optimal solutions.

8.4.2 Equilibrium analysis

For solving problem (8.7) we need the following Hamiltonian H:

$$H(T_{ji}, W_i, \sigma_{ji}, \beta_{ji}, t) = \sum_i \sum_j \iota_i \ln (\beta_{ji} T_{ji}) + \sum_i \sum_j r_j \ln (\sigma_{ji} W_i) + \\ + \sum_i \Theta_i \dot{W}_i + \sum_i \sum_j \Phi_{ji} \dot{T}_{ji} \tag{8.8}$$

as well as the following Lagrangian L:

$$L(T_{ji}, W_i, \sigma_{ji}, \beta_{ji}, t) = W + \mu(\bar{W} - \sum_i W_i) \tag{8.9}$$

where Θ_i and Φ_{ji} represent the costate variables related to the $\dot{W}_i$ and $\dot{T}_{ji}$ constraints, respectively; μ is the shadow price related to the static constraint. Therefore, the first-order (necessary) conditions for the control variables become:

$$\frac{\partial L}{\partial \beta_{ji}} = 0 \\ \frac{\partial L}{\partial \sigma_{ji}} = 0 \tag{8.10}$$

so that the optimal control values are (see (8.8), (8.9) and (8.10)):

$$\beta_{ji} = \frac{\iota_i}{\Theta_i W_i T_{ji}}$$

$$\sigma_{ji} = \frac{-r_j}{\Phi_{ji} W_i T_{ji}} \tag{8.11}$$

From (8.11) we can also derive that the optimal values β_{ji} are a decreasing function of the shadow prices Θ_i (i.e. $\partial\beta_{ji}/\partial\Theta_i < 0$).

Next we find from (8.11) that, since σ_{ji}, T_{ji}, W_i, r_j are positive, the shadow prices Φ_{ji} are negative and also that σ_{ji} is an increasing function of the corresponding shadow price Φ_{ji} (i.e., $\partial\sigma_{ji}/\partial\Phi_{ji} > 0$).

Next, a link between the shadow prices Θ_i and Φ_{ji} can be found. In fact from (8.11) it follows that:

$$\frac{\Phi_{ji}}{\Theta_i} = - \frac{\beta_{ji}}{\sigma_{ji}} \frac{r_j}{\iota_i} \tag{8.12}$$

In other words, it appears that the ratio of shadow prices of the $\dot{W}_i$ and $\dot{T}_{ji}$ constraints is inversely related to the corresponding optimal control variables (i.e., the interaction terms β_{ji} and σ_{ji} in the prey-predator model (8.5) and (8.6)).

Next we will deal with the costate equations. Here the following conditions hold:

$$\dot{\Theta}_i = - \frac{\partial L}{\partial W_i}$$

$$\dot{\Phi}_{ji} = - \frac{\partial L}{\partial T_{ji}} \tag{8.13}$$

or:

$$\dot{\Theta}_i = - \frac{\sum_j r_j}{W_i} - \alpha_i \Theta_i + \Theta_i \sum_j \beta_{ji} T_{ji} - \sum_j \Phi_{ji} \sigma_{ji} T_{ji} + \mu$$

$$\dot{\Phi}_{ji} = - \frac{\iota_i}{T_{ji}} + \phi_i \Phi_{ji} + \Theta_i \beta_{ji} W_i - \Phi_{ji} \sigma_{ji} W_i \tag{8.14}$$

Then, by substituting the optimal values (8.11) into (8.14), we obtain:

$$\dot{\Theta}_i = \frac{\iota_i}{W_i} - \alpha_i \Theta_i + \mu$$
$$\dot{\Phi}_{ji} = \frac{r_j}{T_{ji}} + \Phi_{ji} \phi_i \tag{8.15}$$

It is clear now that a phase diagram analysis of the qualitative properties of the solutions may offer a very interesting analytical tool (see also Chapter 5). This type of analysis with reference to the state variables - costate variables plane is illustrated in Annex 8A. In particular we can see that, by using the approximative linear differential equation system in the neighbourhood of the steady states, we obtain a result which can be classified as 'degenerated centre' (see Annex 6A).

In other words, the emerging equilibrium points seem to belong to a **borderline** case so that no definite conclusion can be drawn about the nature of the steady states. In particular, we may suppose the possibility of bifurcation phenomena depending on the direction of movement in the original non-linear system, most probably correlated with borderline cases of **Hopf** bifurcations (discussed in Section 6.2.1) emerging from the conditions (8A.9), (8A.10) and (8A.11).

In conclusion, our stability analysis has shown for synergetic models of spatial interaction based on dynamic ecological principles (see equations (8.5) and (8.6)) that an optimal control approach poses specific problems, as stable optimal paths cannot be ensured (cf. also Dendrinos and Mullally, 1983). However, it can be interesting to analyze the complex behaviour emerging from these unstable trajectories: in fact by manipulating the parameters different kinds of bifurcation phenomena may emerge, likely based on limit cycles. Furthermore, for systems with dimension $n \geq 3$ optimal trajectories leading to chaotic behaviour can be examined (see Section 6.2.2).

8.4.3 Concluding remarks

In order to conclude our analysis of ecological models in the context of dynamic spatial interaction systems, we will also examine the **symbiosis** hypothesis for the state variables (W_i, T_{ji}) in our optimal control problem (8.7) (i.e., mutual positive spinoffs between inflows and vacant workplaces). It is very easy to see that no drastic change will occur in the stability analysis for this type of system (see Nijkamp and Reggiani, 1990a).

Finally, the case of 'competing species' concerning workplaces and inflows is worth mentioning, i.e., if one species increases in size, the growth rate of the other declines (see, among others, Haken, 1983a, Hirsch and Smale, 1974, Johansson and Nijkamp, 1987, and Serra and Andretta, 1984). The general case of spatial competition in dynamic systems will be extensively treated in the next section. With reference to our optimal control problem (8.7), it can easily be seen that the stability analysis related to the use of competing species will lead to the same conclusions as in the other two cases (see Nijkamp and Reggiani, 1990a). Obviously our particular results depend on the specific shape of the objective function adopted. In fact, it can easily be seen that by using a more general objective function incorporating also different control variables, we obtain at the end a system in four differential equations which is unfortunately very difficult to handle formally and analytically in the stability analysis (although simulation experiments might be meaningful here).

The ecological analysis previously described therefore leads to the possibility of exploring new research directions in non-linear dynamics. In particular a further orientation toward the theory of chaos may be interesting. It has already been shown that a discrete system of a prey-predator type may lead to a complex behaviour with strange attractors (see Peitgen and Richter, 1986). In this context a more thorough ecological analysis in the way described above (with the solution of a borderline unstable equilibrium) may offer new insights into complex dynamics. Furthermore, it is also be interesting to investigate the impact of a chaotic species in a system that in itself is not oscillating (such as the competing species system), in order to verify the strength (in terms of destabilizing effect) of a chaotic pattern. This will be the subject of the next sections.

8.5 Competition Models: Introduction

Competition models are, together with prey-predator models, frequently used in ecology, since the coexistence of two or multiple species in the same habitat is an interesting phenomenon, leading to different types of behaviour, as we will see in this section. Competition models are a slightly different version of the prey-predator model (8.1), since they are formulated (in a continuous form) as follows:

$$\begin{aligned} \dot{x} &= x\,(a - bx - cy) \\ \dot{y} &= y\,(d - ex - fy) \end{aligned} \tag{8.16}$$

where the terms -cy and -ex describe the inhibiting effect of each species x,y on its competitor. Consequently, c and e are the so-called competition coefficients. It should be noted that the competing terms are added to the usual logistic equation. System (8.16) has been studied by several authors starting from Lotka (1925), Volterra (1926), and Gause (1934).

In particular Maynard Smith (1974) analyzed system (8.16) for biological processes, by showing -in case of a non-trivial equilibrum- two cases:

i) a stable equilibrium for $a/b < d/e$ and $d/f < a/c$ (see Figure 8.5);

ii) an unstable equilibrium with both inequalities reversed (see Figure 8.6), in which case either species can become a winner, depending on the initial conditions.

It is important to underline that Maynard Smith (1974) showed that, both for system (8.16) and for a general case of competition between two species, described as follows:

$$\dot{x} = g\,x\,(x, y)$$

$$\dot{y} = h\,y\,(x, y) \tag{8.17}$$

the non-trivial equilibrium - if it exists - is stable or unstable but **non-oscillatory**.

In particular the equilibrium is stable when

$$\frac{\partial g}{\partial x} \cdot \frac{\partial h}{\partial y} > \frac{\partial g}{\partial y} \cdot \frac{\partial h}{\partial x} \tag{8.18}$$

where the differentials $\left(\frac{\partial .}{\partial .}\right)$ are calculated at a non-trivial equilibrium point.

Expression (8.18) shows at the left hand side the 'marginal' inhibiting effects of each species on itself and at the right hand side the effects of each species on its competitor. In other words, following Gilpin and Justice (1972) it is clear that "a necessary and sufficient condition for the stability of a competitive equilibrium is that the product of the intraspecific growths regulation be greater than the product of the interspecific growth regulations" (Maynard Smith, 1974, p. 64).

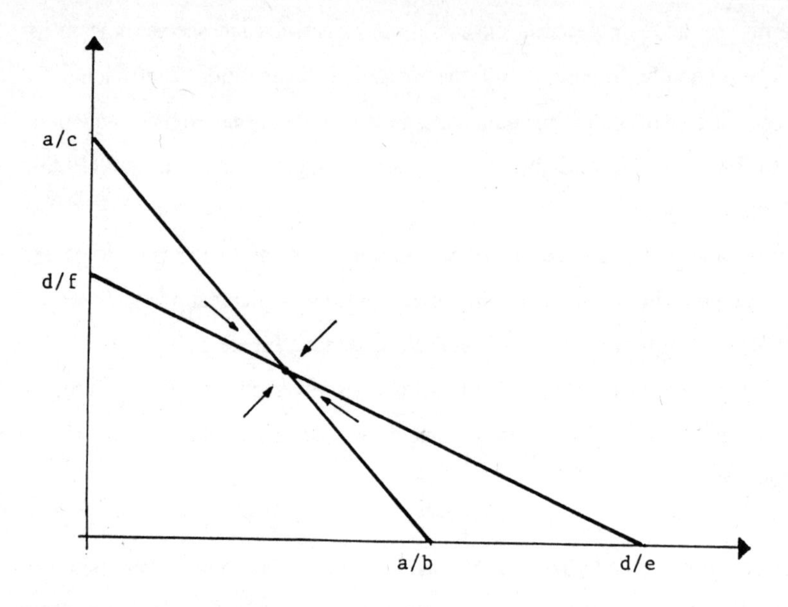

Figure 8.5 Stable Equilibrium in a Competition Between Two Species x and y
y-axis= 1:x; 2:y; x-axis = time

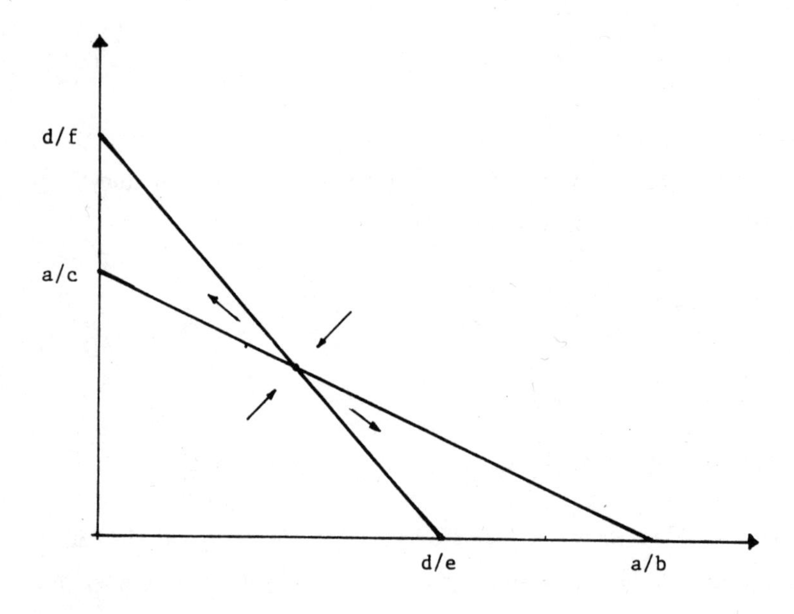

Figure 8.6 Unstable Equilibrium in a Competition Between Two Species x and y
y-axis: 1=x; 2=y; x-axis = time

Condition (8.18) stems directly from the formal expression of (8.16) or (8.17), since by definition an increase in one species reduces the 'velocity of increase' of the other one. This also means that if the two systems have identical needs with the same resources, the more efficient system will eliminate its competitor.

This result has in our analytical framework another important meaning. It hampers the emergence of chaotic behaviour in systems of type (8.16), since a pre-condition for chaos is the emergence of a chain of Hopf bifurcations (at least three) which are based on 'neutral stability' (a limit cycle or oscillatory behaviour) (see Section 6.2).

Obviously, it is interesting to notice that this analysis can also be considered in an spatial economic framework. For example, the competing species can be represented by competing centres or regions, so that the competition coefficients are parameters depending on a relevant significant variable (such as accessibility, gross product, etc.) which could be tested empirically. We can also analyse the case of competing species in discrete time by considering a general system describing competition between species with discrete generations:

$$x_{t+1} = x_t \, p \, (x_t, y_t)$$

$$y_{t+1} = y_t \, q \, (x_t, y_t) \qquad (8.19)$$

Also for the general system of type (8.19) Maynard Smith (1974) showed that the equilibrium - if it exists - is either stable or unstable, but in either case **non-oscillatory**. Moreover he demonstrated the absence of oscillations in the long run. It should be noted however that the same result does not emerge in interactions of a prey-predator type, where also divergent oscillations can emerge.

Starting from these results which can also be transferred and applied to competing centres in a network system, e.g., we will show in the next section how a very particular case of (8.16) (i.e., when the first place (region) follows a chaotic evolution without a direct competition effect on the second one), can indeed show, under certain conditions, the emergence of oscillating (likely chaotic) behaviour.

8.6 Impact of Chaotic Evolution in Spatial Competition

8.6.1 Introduction

In this section, starting from system (8.16) in the context of a spatial network, we will consider the particular case in which a region follows a 'chaotic' evolution of the well-known May type of model. In other words, we suppose that in the first equation of system (8.16) the competition coefficient c measuring the effect to which center y exerts a pressure upon the resources used by x is equal to 0. This hypothesis may be plausible from a spatial-economic viewpoint, as this may reflect a hierarchy in spatial systems. This may imply a situation where higher-order places (or regions) have a decisive influence on lower-order places (or regions), without being influenced by means of feedback effects by lower-order places (or regions). Both Christaller-Lösch systems but also international trade block dominances may be described by such analytical systems.

In particular, we will analyze here the impact of this 'chaotic' region in the case of two competing regions.

8.6.2 The case of two competing regions

8.6.2.1 Equilibrium analysis

Under the above mentioned hypothesis of a 'chaotic' region, the implications for the continuous system (8.16) are as follows:

$$\dot{x} = x\,(a - bx)$$

$$\dot{y} = y\,(d - ex - fy) \qquad (8.20)$$

where: $m = a/b$; $n = d$

We will now analyze the discrete version of system (8.20), since this makes more sense from an economic viewpoint in terms of empirical applications (data are usually only available in discrete form), while also simulation experiments are usually carried out in discrete terms.

The discrete form of system (8.20) is therefore the following (see Annex 8B for the mathematical derivations):

$$x_{t+1} = \lambda\, x_t\,(1 - x_t)$$

$$y_{t+1} = y_t\,(m - ex_t - fy_t) \qquad (8.21)$$

where (see expressions (8B.2), (8B.5) and (8B.9) in Annex 8B):

$$\lambda = a + 1$$
$$m = d + 1 \tag{8.22}$$

It should be noted here that system (8.20) can also be derived on the basis of the prototype model of systemic competition developed by Johansson and Nijkamp (1987) in their analysis of urban and regional development:

$$\dot{x}_i = \delta_i x_i (N_i - x_i - \sum_j \rho_{ij} x_j) - \gamma_i x_i \tag{8.23}$$

where:
x_i = production (or income) of place i
δ_i = entry (expansion) rate of i
γ_i = exit (depreciation) rate of i
N_i = carrying capacity of production level x_i
ρ_{ij} = competition coefficient

Thus in model (8.23) a positive ρ_{ij} reflects competition between different types of production, while a negative ρ_{ij} signifies that sector i is positively stimulated by the growth of the other sectors. Thus model (8.23) may describe a wide variety of episodes, such as exclusion through competition for the same resource, or competition for the same market or synergetic reinforcement.

Transformation of system (8.23), in the two-dimensional case (i = 1,2) (where $x_1 = x$ and $x_2 = y$), leads after some mathematical derivations to system (8.16) (see Nijkamp and Reggiani, 1991b).

Let us now consider in (8.23) the case of one region evolving 'chaotically' thus leading to the particular case of the system described in (8.21).

The equilibrium analysis related to system (8.21) then shows that two fixed points exist (see Annex 8C): a trivial one P_1 (0,0) and a non-trivial one P_2 $[(\lambda - 1)/\lambda; (m\lambda - e\lambda + e - \lambda)/\lambda f]$. The stability analysis for P_1 shows unstable behaviour for $\lambda > 1$ and $m > 1$. If we now examine the stability of point P_2, we get the critical value (see equation (8C.9) in Annex 8C):

$$m^* = (e\lambda^2 + 2\lambda^2 - 3\lambda - 3e\lambda + 2e) / (\lambda^2 - 2\lambda) \tag{8.24}$$

at which a Hopf bifurcation (i.e., a bifurcation of a fixed point into a closed orbit) occurs. This implies that for $m > m^*$ the fixed point P_2 becomes unstable, with the possibility of oscillations. In other words this result shows that **if** one equation in the general system (8.19) is reduced to an equation of a May type (leading to chaos), we get in the whole system **the possibility of oscillating behaviour** based on Hopf bifurcations, and hence unpredictable movements. This result is indeed interesting since it shows that a 'chaotic' evolution can introduce oscillations in a system which in itself is non-oscillatory as the one defined in (8.19).

The above results can also be illustrated graphically by means of simulation experiments. In particular, by fixing the parameter e, we can analyze the dynamics of system (8.21) under particular values of the growth parameter λ (leading to chaos). For this purpose we will investigate analytically from (8.24) the values of m^* leading to instability (see Table 8.1).

Table 8.1 Values of Some Growth Parameters Leading to a Hopf Bifurcation

m^*	λ	e
3.45	3.1	0.8
3.20	3.6	0.8
3.12	3.9	0.8
3.10	3.99	0.8

It should be noted that the values of λ in Table 8.1 are the ones leading to irregular behaviour (λ= 3.1; λ = 3.6) and to chaotic behaviour (λ = 3.9; λ = 3.99) in a May equation (see also Section 6.2.2).

Furthermore we can also point out the relevance of the competition coefficient e, since it can be seen that it is directly proportional to the critical value m^*. In particular, by decreasing the competition effect captured by the parameter e, also the instability of system (8.21) occurs at a lower value of m^* (see, for example Table 8.2, where some values of e and m^* are illustrated by fixing the 'chaotic' value of λ = 3.9).

Table 8.2 Relationships between Competition Coefficient e and Growth Parameter m^*, for fixed value $\lambda = 3.9$.

e	0.8	0.69	0.5
m*	3.12	3.04	2.90

8.6.2.2 Simulation experiments

For the first simulation (see Figure 8.7) we have assumed the following parameter values:

$$m = 4.2 > m^* \qquad \lambda = 3.1 \qquad e = 0.8 \qquad f = 1$$

with the initial conditions:

$$x = y = 0.1$$

Figure 8.7 shows a periodic behaviour, which may likely - depending on the value of $\lambda = 3.1$ - lead to cycles. It is interesting to note that the value of parameter f has no impact on the emergence of unstable behaviour (see also equation (8.24)).

In the following simulations we will consider values of λ that display in the corresponding May equation both irregular behaviour (for example, for $\lambda = 3.6$) and chaotic behaviour (for example, for $\lambda = 3.9$). Then we will consider two cases:

$$m < m^* \text{ and } m > m^*$$

Therefore, for the second simulation (Figure 8.8) we will assume:

$$m = 1.5 < m^* \qquad \lambda = 3.6 \qquad e = 0.8$$

while for the third simulation (Figure 8.9) we will assume:

$$m = 1.5 < m^* \qquad \lambda = 3.9 \qquad e = 0.8$$

with the same initial conditions:

$$x = y = 0.1$$

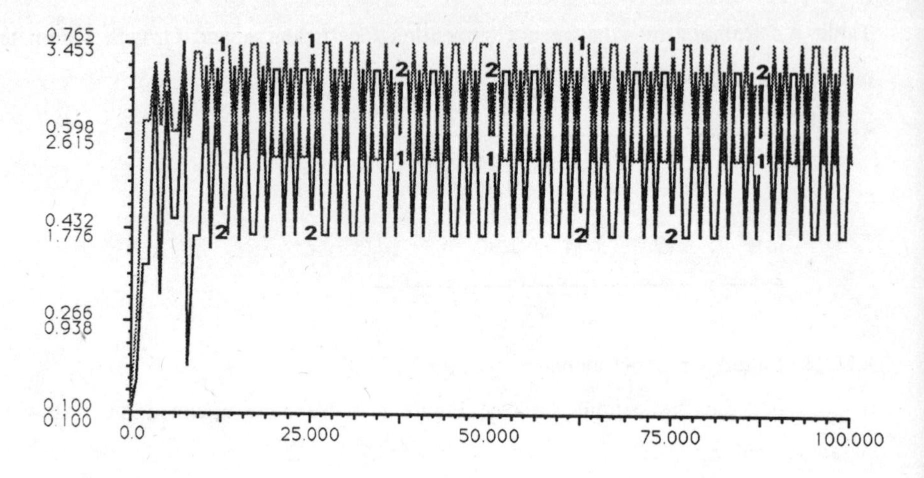

Figure 8.7 Regular Oscillatory Behaviour for the Two Competing Regions x and y
y-axis= 1:x; 2:y; x-axis=time

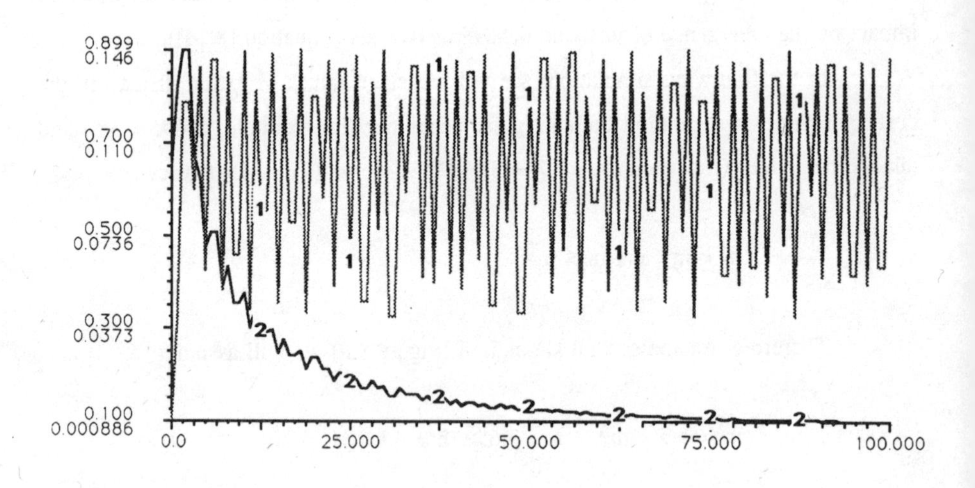

Figure 8.8 Oscillatory Behaviour for Region x and Stable Behaviour (in the Long Run) for Region y (for m < m* and $\lambda = 3.6$)
y-axis= 1:x; 2:y; x-axis=time

Figure 8.8 shows that by keeping the growth parameter m below the critical value m* the related equation shows stability in the long run. The same happens even if we assume a 'chaotic' value of λ (λ= 3.9) (see Figure 8.9). It turns out that the competing region y is more oscillatory in the short run; however, in the long run it reaches stability (by being eliminated by region x).

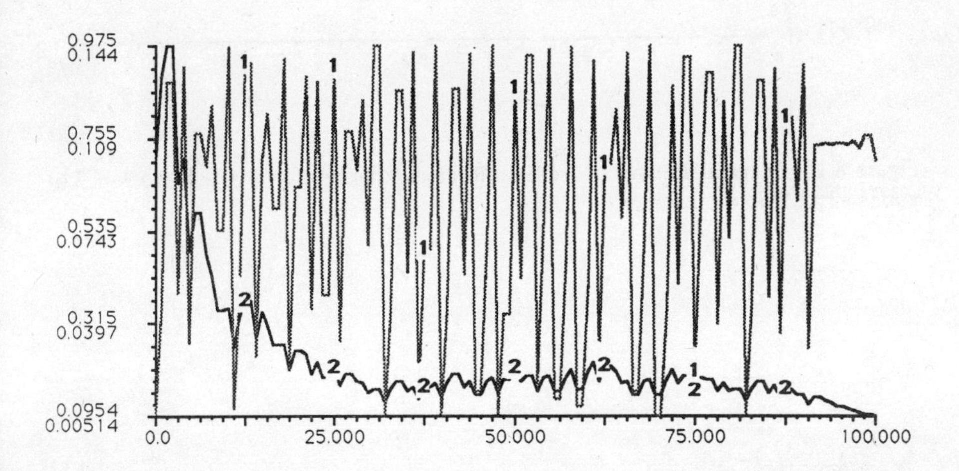

Figure 8.9 Again Oscillatory Behaviour for Region x and Stable Behaviour (in the Long Run) for Region y (for m < m* and λ = 3.9 ('chaotic' value))
y-axis= 1:x; 2:y; x-axis=time

However, if we consider a value of the growth parameter m beyond the critical value m* (by maintaining the same values for λ and e considered in Figures 8.8 and 8.9), we can clearly observe a destabilization of region y.

In particular, Figure 8.10 shows the emerging irregular behaviour in the evolution of region y for the following parameter values:

$$m = 3.5 > m^* \qquad \lambda = 3.6 \qquad e = 0.8$$

while Figure 8.11 displays an even more irregular pattern, due to the increased value of λ (λ = 3.9, i.e. a 'chaotic' value) in combination with a value of m = 3.5 > m* (see equation (8.24)) (by keeping e = 0.8 and the same initial conditions x = y = 0.1).

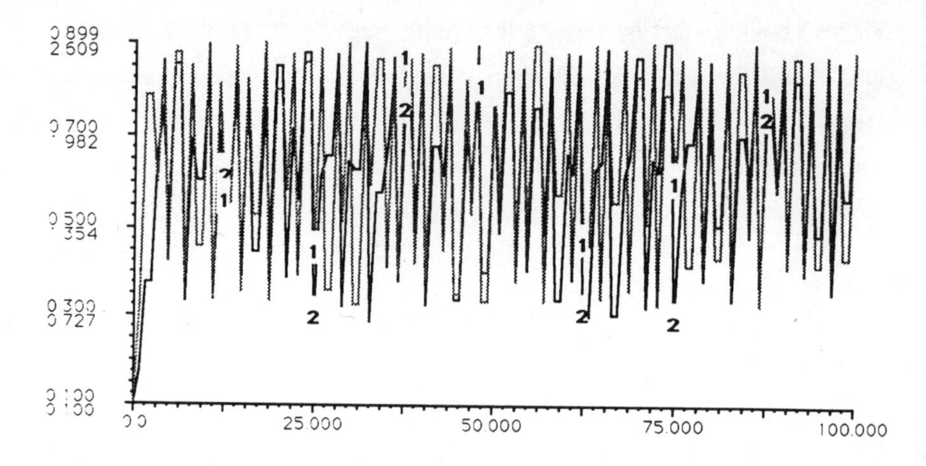

Figure 8.10 Irregular Behaviour for Both Regions x and y (for m > m* and λ = 3.6)
y-axis= 1:x; 2:y; x-axis=time

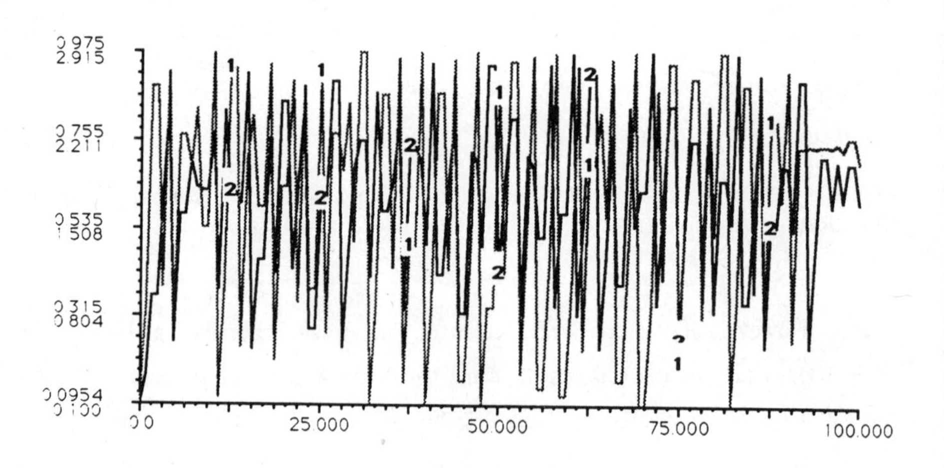

Figure 8.11 Again Irregular Behaviour for Both Regions x and y (for m > m* and λ = 3.9 ('chaotic' value))
y-axis= 1:x; 2:y; x-axis=time

8.6.3 Concluding remarks

In conclusion, the analysis carried out in the case of two competing regions, one of which is 'chaotic', shows the following result:

a) **stability** occurs at **low values** of the growth rates;

b) a 'chaotic' region becomes **unstable** with increasing growth rates, although it does not completely destabilize (in the long run) the competitive region, unless the growth rate of the competing region exceeds the critical value m* at which a Hopf bifurcation occurs.

Furthermore, the equilibrium analysis related to this particular case shows the possibility of oscillating behaviour based on a Hopf bifurcation. Thus our analysis broadened Maynard Smith's (1974) statement which defines the absence of oscillating behaviour in a general two-dimensional spatial competition system. This illustrates the relevance of the growth rates of two competing regions, since the combination of these two factors may lead to instability. Thus the importance of the speed of change of these parameters in real systems again becomes clear (see also Chapters 6 and 7), since a variation in these parameters may lead to a switch from stable to unstable behaviour.

Finally, this result provides fruitful insights into the actual mechanisms of competition and suggests new directions of research which still have to be undertaken in this context:

A) the formulation of a theory on interactive competition coefficients (virtually non-existent); for example, a study of decisive variables on which the competition coefficients are dependent (e.g., density of population, accessibility, environmental-technological conditions, etc.)

B) the definition of the 'real' domain of the relevant parameter space (e.g., an extensive stable domain or not) in order to test the dynamic (in)stability for both the predictable and unpredictable systems at hand

C) the study of the relationships between stability and complexity; in other words if stability increases by increasing the number of competing regions as it happens in ecology (see Maynard Smith, 1974)

D) the development of empirical research by gathering data on the speed and spread of change of competition coefficients as well as intrinsic growth rates.

8.7 Epilogue

In this chapter synergetic models of spatial interaction (i.e., ecologically-based models such as prey-predator, symbiosis and competing species) have been analyzed. In particular the first part of this chapter has shown that these synergetic models may display a borderline stability in the context of an optimal control problem for a spatial interaction analysis (related here to the labour market). In other words, stable optimal paths cannot be ensured. This observation leads to the issue of investigating spatial interaction systems in the context of states of disequilibrium and unpredictability. Following these lines, the second part of this chapter has been devoted to the analysis of the impact of a discrete logistic growth (leading to chaotic, endogenous fluctuations) in a more general system, such as the spatial competition model which in itself is non-oscillatory.

This analysis, which has been carried out for the case of two competing regions, shows on the one hand the relevance of a chaotic pattern, and on the other hand the relevance of the growth rates of competing regions.

In particular we may underline here that chaotic behaviour in a region emerges only in a certain limited range of its growth rate and that it influences (in terms of irregularity) the competing region only if the related intrinsic growth rate (depending also on the interregional competition coefficient) exceeds a critical value, at which a Hopf bifurcation occurs.

Thus, the relevance of the logistic discrete growth in this type of spatial dynamic models has become evident. In the next chapter we will use and interpret the logistic growth as the basis of many dynamic non-linear models adopted so far in economics and in regional science. Consequently, also the possibility of chaotic movements seems to emerge in all these models, under particular conditions of the parameters and initial conditions.

Annex 8A

Stability Solutions for an Optimal Control Prey-Predator Problem

Given the steady state solutions (8.11) of the ecological optimal control model (8.7), we now wish to infer a number of conclusions regarding the nature of these steady states, with reference to the qualitative analysis in the state variables - costate variables phase.

Therefore, by substituting the optimal values (8.11) in the differential equations of (8.7) we obtain:

$$\begin{aligned} \dot{W}_i &= \alpha_i W_i - \frac{\iota_i}{\Theta_i} \\ \dot{T}_{ji} &= -\phi_i T_{ji} - \frac{r_j}{\Phi_{ji}} \end{aligned} \tag{8A.1}$$

By considering now the system (8.15) derived from the transversality conditions we get the following four-dimensional system:

$$\begin{aligned} \dot{W}_i &= \alpha_i W_i - \frac{\iota_i}{\Theta_i} \\ \dot{\Theta}_i &= \frac{\iota_i}{W_i} - \alpha_i \Theta_i + \mu \\ \dot{T}_{ji} &= -\phi_i T_{ji} - \frac{r_j}{\Phi_{ji}} \\ \dot{\Phi}_{ji} &= \frac{r_j}{T_{ji}} + \Phi_{ji} \phi_i \end{aligned} \tag{8A.2}$$

It is interesting to see that system (8A.2) can easily be subdivided into two **independent** subsystems as follows:

$$\begin{aligned} \dot{T}_{ji} &= -\phi_i T_{ji} - \frac{r_j}{\Phi_{ji}} \\ \dot{\Phi}_{ji} &= \frac{r_j}{T_{ji}} + \Phi_{ji} \phi_i \end{aligned} \tag{8A.3}$$

and:

$$\dot{W}_i = \alpha_i W_i - \frac{\iota_i}{\Theta_i}$$

$$\dot{\Theta}_i = \frac{\iota_i}{W_i} - \alpha_i \Theta_i + \mu \qquad (8A.4)$$

We shall first examine system (8A.3), defined in the plane (T_{ji}, Φ_{ji}). It is evident that the equilibrium points of the two differential equations $\dot{T}_{ji} = \dot{\Phi}_{ji} = 0$ describe the same hyperbola (see Figure 8A.1):

$$T_{ji} = \frac{-r_j}{\Phi_{ji}\,\phi_i} \qquad (8A.5)$$

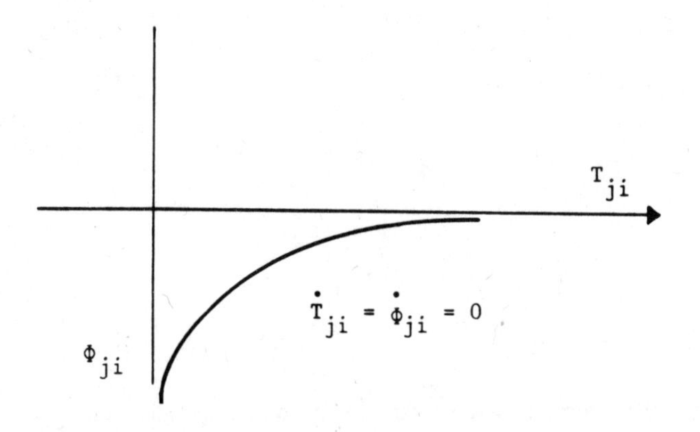

Figure 8A.1 Optimal Paths Consistent with the Differential Equation (8A.3)

This means that **all points** of the hyperbola (8A.5) are equilibrium points.

To catalogue the nature of these steady states, we will take the linear terms of Taylor series expansion at the right hand side of system (8A.3) (around all steady states $T_{ji,s}$, $\Phi_{ji,s}$) to obtain the approximate linear differential equation system:

$$\dot{T}_{ji} = - \phi_i (T_{ji} - T_{ji,s}) + \frac{r_j}{\Phi^2_{ji,s}} (\Phi_{ji} - \Phi_{ji,s})$$

$$\dot{\Phi}_{ji} = - \frac{r_j}{T^2_{ji,s}} (T_{ji} - T_{ji,s}) + \phi_i (\Phi_{ji} - \Phi_{ji,s}) \tag{8A.6}$$

The characteristic equation of (8A.6) with eigenvalues K (= K_1, K_2) is the following:

$$\begin{vmatrix} - \phi_i - K & \dfrac{+ r_j}{\Phi^2_{ji,s}} \\ \dfrac{- r_j}{T^2_{ji,s}} & + \phi_i - K \end{vmatrix} = 0 \tag{8A.7}$$

or:

$$K^2 - pK + q = 0 \tag{8A.8}$$

with:

$$p = K_1 + K_2 = 0 \tag{8A.9}$$

$$q = K_1 K_2 = - \phi_i^2 + \frac{r_j^2}{T^2_{ji,s} \Phi^2_{ji,s}} \tag{8A.10}$$

It is easily seen that from (8A.5) we can derive:

$$\phi_i^2 = \frac{r_j^2}{T_{ji,s}^2 \; \Phi_{ji,s}^2} \tag{8A.11}$$

It is clear that q, defined in (8A.10) is equal to zero. Since $q = p = 0$, it is evident that $K_1 = K_2 = 0$. This case is in the classification of singularities an example of a degenerate case, i.e., a borderline case (see Annex 6A).

The previous analysis is also closely connected to so-called extreme steady states, which provide a stationary solution of the canonical equations for the state and co-state variables if $\tilde{T}$ -> ∞ (see Balducci et al., 1984, Haurie and Leitmann, 1984, and Ricci, 1985).

Finally we have to consider the system defined in the plane (W_i, Θ_i), i.e.,:

$$\begin{aligned} \dot{W}_i &= \alpha_i W_i - \frac{\iota_i}{\Theta_i} \\ \dot{\Theta}_i &= \frac{\iota_i}{W_i} - \alpha_i \Theta_i + \mu \end{aligned} \tag{8A.12}$$

The optimal paths are defined for $\dot{W}_i = \dot{\Theta}_i = 0$, i.e.,

$$W_i = \frac{\iota_i}{\Theta_i \, \alpha_i} \tag{8A.13}$$

and

$$\Theta_i = \frac{\iota_i}{W_i \, \alpha_i} + \frac{\mu}{\alpha_i} \tag{8A.14}$$

It is evident that (8A.13) and (8A.14) have no common intersection points unless the condition $\mu = 0$ holds. This case, which means that the additivity condition $\bar{W}$ plays no role in the optimal control problem (8.7), reduces again (8A.13) and (8A.14) to two coinciding hyperbolas of equilibrium points.

Consequently, linearization of system (8A.12) will lead to the same conclusions about the nature of these equilibrium points found previously, i.e., a **borderline** stability.

Annex 8B

Transformation of a Continuous System into a Discrete System

In this annex we will show how the continuous system (8.20) can be approximated in a discrete form.

For the sake of simplicity we will write the first equation of system (8.20) in the following form:

$$\dot{x} = b\,x\,(n - x) \tag{8B.1}$$

where:

$$n = a/b \tag{8B.2}$$

We can now approximate equation (8B.1) in discrete time by considering a unit time period as follows:

$$x_{t+1} - x_t = bx_t\,(n - x_t) \tag{8B.3}$$

which can also be written as:

$$x_{t+1} = \lambda x_t\,(1 - \frac{\lambda - 1}{\lambda n}\,x_t) \tag{8B.4}$$

where:

$$\lambda = bn + 1 \tag{8B.5}$$

It is now clear that if we make the transformation (see also Nijkamp and Reggiani, 1990b, and Wilson, 1981):

$$x_t^* = x_t\,(\lambda - 1)\,/\,\lambda n \tag{8B.6}$$

equation (8B.4) can be written in the canonical form:

$$x^*_{t+1} = \lambda\,(1 - x^*_t) \tag{8B.7}$$

For the sake of simplicity we have rewritten equation (8B.7) in system (8.21) in terms of x instead of x*, since this is a translation of coordinates without any impact on the behaviour of the system.

Next, we will consider the approximation in discrete terms of the second equation in (8.20) as follows:

$$y_{t+1} - y_t = y_t\,(d - ex_t - y_t) \tag{8B.8}$$

or

$$y_{t+1} = y_t\,(d + 1 - ex_t - y_t) \tag{8B.9}$$

Consequently, equations (8B.7) and (8B.9) constitute in system (8.21) the difference version of system (8.20).

Annex 8C

Stability Analysis for a Particular Competing System

In this annex we will analyse the stability properties of system (8.21). For the ease of reference we will write again system (8.21)

$$x_{t+1} = x_t\,(\lambda - \lambda x_t)$$
$$y_{t+1} = y_t\,(m - ex_t - fy_t) \tag{8C.1}$$

It is easy to find the fixed points for which $x_{t+1} = x_t$ and $y_{t+1} = y_t$. These are P_1 (0,0) and $P_2[(\lambda - 1)/\lambda; (m\lambda - e\lambda + e - \lambda)/\lambda f]$.

The local behaviour of the map (8C.1) at its fixed points P_1, P_2 is governed by its local linearization for which the Jacobian J:

$$J = \begin{pmatrix} -2\lambda x^* & 0 \\ -ey^* & m-ex^*-2y^* \end{pmatrix} \tag{8C.2}$$

taken at the fixed point P (x*, y*) is the corresponding matrix.

Consequently we get:

$$J(P_1) = \begin{pmatrix} \lambda & 0 \\ 0 & m \end{pmatrix} \tag{8C.3}$$

In other words the multipliers of P_1 (i.e. the eigenvalues K_1 and K_2 of (8C.3)) are:

$$K_1 = \lambda$$
$$K_2 = m \tag{8C.4}$$

It is clear that the trivial fixed point P_1 (0,0) shows a rich spectrum of stable/unstable behaviour according to the values of growth rates λ and m of regions x and y, respectively.

In particular we can underline the following results (for the basic notations related to the stability of discrete-time system see, e.g., Lauwerier, 1986, and Lorenz, 1989):

a) if $\lambda < 1$ and $m < 1$, fixed point P_1 is stable;

b) if $\lambda < 1$ and $m > 1$, fixed point P_1 is a saddle;

c) if $\lambda > 1$ and $m > 1$, fixed point P_1 is a repelling node;

d) if $\lambda = 1/m$ a Hopf bifurcation occurs.

We will now analyse the Jacobian (8C.2) taken at P_2:

$$J(P_2) = \begin{pmatrix} 2-\lambda & 0 \\ \frac{-e}{\lambda f}(m\lambda - e\lambda + e - \lambda) & \frac{e\lambda - m\lambda - e + 2\lambda}{\lambda} \end{pmatrix} \tag{8C.5}$$

Here we consider the trace Tr of (B.5) as well as its determinant Det:

$$\text{Tr } J(P_2) = J_{11} + J_{22} = \frac{e\lambda - m\lambda - e - 4\lambda - \lambda^2}{\lambda} \tag{8C.6}$$

$$\text{Det } J(P_2) = J_{11} \cdot J_{22} = \frac{(2-\lambda)(e\lambda - m\lambda - e + 2\lambda)}{\lambda} \tag{8C.7}$$

If follows from Ruelle-Takens theorem (1971) that a **Hopf bifurcation occurs** at $m = m^*$ such that $\text{Det } J(P_2|_{m=m^*}) = 1$, i.e.:

$$(2-\lambda)(e\lambda - m^*\lambda - e + 2\lambda) = \lambda \tag{8C.8}$$

or

$$m^* = \frac{e\lambda^2 + 2\lambda^2 - 3\lambda - 3e\lambda + 2e}{\lambda^2 - 2\lambda} > 0 \tag{8C.9}$$

Thus this implies that for $m > m^*$ the eigenvalues have a modulus larger than unity, i.e. the fixed point becomes unstable.

It should be noted that the Hopf bifurcation in discrete-time models has rarely been applied in economics, although there are a few exceptions (see, e.g., Cugno and Montrucchio, 1984b, Farmer, 1986, Lorenz, 1989, and Reichlin, 1985). Our interest in finding the parameter values m^* leading to a Hopf bifurcation is caused by the fact that a Hopf bifurcation is one of the possible routes to chaos (see Schuster, 1988 and Section 6.2.1).

CHAPTER 9

RETROSPECT AND PROSPECT

9.1 Retrospect

In this study various research directions and models centering around SIMs have been analyzed. In the first part of the book the attention has been focused on static developments in SIMs, while the second part has been devoted to an analysis of dynamic aspects of SIMs with particular emphasis on chaos theory.

As a synthesis, the main results emerging from the first part (i.e., the static analysis) of the present study can be summarized as follows:

- SIMs have various roots. The most common specification of a SIM originates from **gravity theory** (see Chapter 1).
- A second macro-oriented approach from which SIMs can emerge is the entropy concept. Even though the entropy model has its roots in statistical mechanisms, the entropy concept can be interpreted in terms of a **generalized cost function** for spatial interaction behaviour, thus offering a macro-behavioural context to SIMs (see Chapter 2).
- A further utility background of entropy is offered by various **programming (or optimization) models**, since entropy can be regarded as a specific type of these latter models. Consequently, the family of SIMs can be derived from different formulations of an entropy (or utility) maximizing macro-approach and hence viewed as an optimum system's solution. Once again, entropy theory offers a result that is compatible with a macro-behavioural interpretation of spatial interaction (see Chapter 3).
- SIMs can also emerge from the maximization of **micro-economic (deterministic or random) choice theory**. In particular, the latter approach illustrates that SIMs are formally analogous to DCMs; consequently, depending on the type of available data (aggregate or disaggregate) SIMs can provide a similar behavioural background as DCMs. In particular, a singly-constrained SIM can be considered equivalent to an MNL model thus embedding also the limits inherent in MNL models such as the IIA property. Only a doubly-constrained SIM - and more generally the Alonso type of model - with a sequential choice process may be associated with a nested

MNL model, thus overcoming the IIA axiom. It then follows that entropy can also be interpreted as a measure of interaction between economic individuals and consequently as a social/collective utility function (including cost elements) (see Chapter 4).

Furthermore, due to the variety in formulation, SIMs are suitable to describe and explain various behavioural processes underlying individual or group choices. For this purpose much attention has to be paid to a proper specification of the utility function of actors in each given geographical context. This is illustrated in an empirical application related to the field of travel decisions (see again Chapter 4).

Next, the second part of the present study (i.e., dynamic analysis) has led to the following main results:

- The formal parallel between SIMs and DCMs can also be extended to a dynamic and stochastic context. Moreover, a SIM can be shown to be the optimal solution of a **dynamic entropy optimal control problem.** Consequently, the interrelations between entropy and SIMs exist also in a dynamic framework. This indicates that dynamic SIMs are able to capture also the evolutionary patterns of a dynamic interaction system and that a dynamic entropy can be viewed as a cumulative utility function concerned with generalized cost minimization (see Chapter 5).

  SIMs also have the possibility of explaining the interaction activities in a **stochastic** dynamic framework. In fact, a stochastic (doubly-constrained) SIM can be shown to emerge as an optimum system's solution from a dynamic entropy optimal control problem subject to random disturbances. Furthermore, a singly-constrained SIM shows a structural stability even in the presence of such small perturbations, as was shown in an illustration of **catastrophe behaviour** in the field of mobility (see again Chapter 5).
- The presence and relevance of stochastic (exogenous) fluctuations (according to which the usual form of a SIM varies) leads to the issue of analyzing SIMs in the context of states of disequilibrium, causality and unpredictability; in other words in the context of a **chaotic behaviour**. The discovery of 'chaos' seems to have created a new paradigm in scientific modelling. Chaos models in economic and social sciences however, are often theoretical, given the lack of available data (see Chapter 6).

In particular, spatial models in the framework of chaos theory show that irregular dynamic behaviour is a possibility depending on initial conditions and on critical parameter values. Consequently, much attention has to be paid to the speed of change of the parameters, since some critical parameter values can lead to a chaotic movement with unfeasible values for some economic variables. This has been shown in our study by considering a generalized Lorenz model applied to the evolution of a dynamic urban spatial interaction system (see again Chapter 6).

- SIMs have the possibility of showing chaotic features in their MNL formulation, in particular for high values of the corresponding **marginal utility function**. This result opens the issue of compatibility between rational and chaotic behaviour (see Chapter 7). Furthermore, a delayed SIM can exhibit chaotic trajectories leading to a **strange attractor** of the Hénon type (and hence with a fractal character), while even a stable spatial interaction system can be destabilized by time lag effects. In particular, multiple lags in a dynamic SIM may produce a complex behaviour (under specific conditions of the utility function) which falls however in an even smaller region of the parameter space as the number of delays increases (see again Chapter 7).
- **Synergetic models** of spatial interaction (i.e., ecologically-based models such as prey-predator, symbiosis, competing species) may show a borderline stability in the context of an optimal control problem for a spatial interaction problem, as was illustrated in a model on an urban labour market. In other words, stable optimal paths cannot be ensured (see Chapter 8). This observation leads to the issue of analyzing synergetic models (such as the competition models) in the context of chaos. In particular the impact of a 'chaotic' species on a system which in itself is non-oscillatory deserves thorough analysis since the chaotic evolution in a region influences, in terms of irregularity, the competing region, under some critical conditions of the related intrinsic growth rate. Thus the relevance of the logistic growth (leading to a chaotic behaviour) in this type of spatial model becomes also obvious (see again Chapter 8).

The previous results illustrate once more the relevance of SIMs in analyzing human spatial behaviour in their pluriform aspects (static, dynamic, stochastic, chaotic). At the end of this exposition, however, it should be noted that SIMs have not been developed in isolation from others. They follow the quantitative research tradition in regional science and transportation science, and can be linked to other models developed in the past decades. Therefore, it is opportune to view the development of SIMs also against the background of other models developed in the area of regional and transportation sciences. In particular, we may refer here to modelling results which have been brought about by scientific research in the sixties and seventies in the following fields:

- the design and use of a large number of **equilibrium-type spatial economic and transportation models**, mostly based on **static** SIMs of the gravity type and widely applied in various research fields in the U.S.A. and Europe (see Chapter 1)
- the development of network equilibrium models by means of **mathematical programming techniques**, mainly based on Wardrop's user optimized principle and its equivalent minimization problem formulated by Beckmann et al. (1956) (for a review, see among others, Florian and Spiess, 1982 Friesz, 1985, and Holden, 1989).

It should be noted that the above mentioned tools mainly deal with the **demand** side in a spatial system. However, it has also been pointed out (see Florian and Gaudry, 1983) that a phenomenon designated as supply at one particular level of a spatial system may become a demand component at another level. Consequently, relationships among different levels can be considered as input-output interactions, so that a precise distinction between supply and demand side can be made only at one particular level of the system. It turns out that often the same tools can be used in both demand and supply side analysis (cf. for example, the concept of accessibility derived from spatial interaction analysis and applied to infrastructure systems; see Rietveld, 1989). In particular SIMs can be considered as input for network equilibrium models (see, for example Sen, 1985) or as a component of models which combine interaction, network equilibrium and other travel or locational choices (see, for example, Batten and Boyce, 1986).

However, most models developed in the sixties and seventies are still static/deterministic equilibrium models and do not consider the time paths followed by the spatial system components as well as the uncertainty of the system and its network (cf. the special issue of Transportation Research edited by Boyce, 1985).

Consequently, the eighties have shown (besides further improvements and refinements of the two above mentioned streams of research):

- much more emphasis on the analysis of **individual motives and on the impact of micro behaviour** on the functioning of spatial systems (in view of the need to a better understanding of the complexity of spatial systems) (see the first part (statics) of this study)
- a strong interest in theoretical dimensions of **dynamic modelling** in order to incorporate newly emerging relevant aspects of system dynamics, such as slow and fast dynamics, uncertainty, bifurcations, catastrophe changes, chaotic behaviour, fractal structures, etc. (see the second part (dynamics) of this study)

Consequently, in order to satisfy these analytical and planning needs, a wide variety of new **behavioural** approaches, mainly based on DCMs and Dynamic Models (DMs) has recently emerged at both the supply and demand side of spatial analysis, inter alia by focusing the attention on behavioural foundations and the level of analysis in spatial systems (macro-micro/meso). Since the link between MNL models (and hence SIMs) and other models pertaining to the whole set of choice behaviour modelling - such as models under conditions of uncertainty - is a research field which still needs further investigation, we will in the next section focus our attention on the broad category of DMs developed so far. In particular we will point out a common similarity in these DMs - despite their different theoretical sources -, viz. their close association - under particular conditions - with dynamic SIMs.

9.2 Typology of Dynamic Spatial Interaction Models

9.2.1 Introduction

In this section, we will give a broader overview of dynamic models related to SIMs in order to show once more also the potential of SIMs in relation to the scale of analysis used in various applications.

As noted in Section 9.1, the last decade has shown a boom in the interest in the development of both behavioural and dynamic models. It is generally expected that such models are capable to describe and represent the behavioural mechanisms underlying the evolutionary changes in complex spatial systems. Consequently, a wide variety of multi-temporal or dynamic spatial models has arisen in the past decade with the aim of providing a stronger and more useful analytical support to planning processes than conventional static tools (such as static SIMs, linear programming, etc.). In this context it is plausible that **SIMs will remain a focal point of analysis**, since they can deal with the complicated and interwoven pattern of human activities in space and time.

For this purpose, we first refer to an important research question which largely concerns the applicability of these dynamic models in relation to the scale of analysis at which various operational developments of these models are taking place. In particular, this important field of reflection concerns the advantages and disadvantages related to the use of macro-meso-micro approaches.

On the one hand, it is evident that aggregate representations may become extremely cumbersome and inefficient when it is necessary to represent complex systems, especially when there is considerable heterogeneity amongst actors in those systems (see Clarke and Wilson, 1986). On the other hand, it is clear that the problems of data availability and computational processing requirements are often in contrast with the need to use a micro-oriented approach. Moreover, the response of a population in aggregated models does not always correspond to an aggregation of the individual responses obtained from a micro model, so that it seems evident that the phenomena being studied require a careful consideration as regards the nature of their level of analysis (see again Clarke and Wilson, 1986). This problem has also been treated in analytical attempts focusing the attention on the interdependencies between micro- and macro-responses, which also depend on the interaction between demand and supply.

In our context of a more satisfactory methodological interpretation of the evolution of SIMs in spatial interaction analysis, we will now briefly typify some main groups of dynamic approaches which have gained much attention in the past decade and have also been mentioned in this study from the viewpoint of the level of analysis, viz. macro and meso-micro approaches.

9.2.2 Macro-dynamic approaches

Several dynamic models of spatial structure and evolution have in recent years been developed at a macro level. We will give here a few illustrations. An example is the model developed by Allen et al. (1978) (see Chapter 5), in which the evolutionary growth of zonal activities is assumed to follow a logistic pattern. Allen et al. design a model which gives a comprehensive representation of urban activities such as employment and residential population. A major finding in this model is that random fluctuations (e.g., changes due to infrastructure constructions) may alter the related urban evolution.

Another dynamic model of the logistic type is the one developed by Harris and Wilson (1978) and Wilson (1981) (see Chapter 5). In this case the standard static SIM for activity allocation has been embedded into a dynamic evolutionary framework, again of a logistic type. Bifurcations and catastrophe behaviour emerge from this model, depending on particular values of the parameters. Obviously owing to this logistic structure also oscillations, cycles and chaotic movements may occur (see Chapter 7).

These two important models have induced a wide spread production of related models both from theoretical and empirical viewpoints, not only in a deterministic but also in a stochastic framework (see also Nijkamp and Reggiani, 1988c and Chapter 5). However, it should also be noted that the above mentioned two models primarily focus on the supply side, without clear dynamic equations for the demand side.

Another stream of research at the macro-level is the series of models based on ecological dynamics of the Volterra-Lotka type (see Chapter 8); in this formulation of interacting biological species, each species is characterized by a birth-death process of the logistic type. Recent publications on this issue show the integration between ecological models and optimal control models (Nijkamp and Reggiani, 1990a), between ecological models and random fluctuations of a white noise type (Campisi, 1986) or between ecological models, spatial interaction models of a gravity type and turbulence (Dendrinos, 1988, and Dendrinos and Sonis, 1990).

Obviously, since a Volterra-Lotka system is a system of interrelated equations, we get by necessity here interaction mechanisms of supply and demand. Furthermore, given the related logistic form, it is also here again possible to find - for critical parameter values - oscillations and complex behaviour.

The last group of macro approaches can be found in the area represented by models based on optimal control approaches or dynamic programming analysis. Also here different forms of equilibrium/ disequibrium may emerge (e.g., saddle points, borderline stability) which show the possibility of unstable motions (see Chapters 5 and 8).

As a synthesis we may conclude that a common trend in these groups of macro-approaches is the development of models that are able to exhibit (under certain conditions) complex or chaotic behaviour and hence outcomes which are hardly foreseeable by modellers and planners. This **lack of predictability of future events** is clearly also a major concern in spatial interaction planning and reflects essentially the beginning of a new phase in the research of spatial interaction analysis. Thus another intriguing research problem is emerging, i.e., the relevance of critical parameter values, such as their speed of change in a geographic or planning context in order to understand whether the system at hand is moving towards a predictable or complex (unpredictable) evolution.

9.2.3 Micro-meso dynamic approaches

In this subsection we will briefly pay attention to the considerable body of literature based on micro-oriented models (see Clarke and Wilson, 1986). Given the above mentioned drawbacks related to a macro-approach, a mixture of aggregate dynamic models in conjunction with micro-simulation (the micro-meso approach) has recently been advocated and adopted for various spatial applications (see also Birkin and Clarke, 1983). In this way also an integration between demand and supply results is possible. In other words, micro-meso dynamic approaches utilize individual data in conjunction with aggregate equations.

Another interesting micro-meso approach is the well-known logit model, based on a micro-economic foundation. It has recently been shown (see Nijkamp and Reggiani, 1990b, 1990d, and Chapter 7) that the growth over time of people choosing such alternatives as travel choice mode, destination, etc. according to a logit process follows again a logistic pattern embedded in a general system of the Volterra-Lotka type. Such developments can also lead to a chaotic or irregular behaviour for particular values of the utility function. The most important consequence of this result is the **link between DCMs** - and hence the equivalent SIMs - **and a logistic formulation** of these models. Since most of the models referred to end up with a logistic shape, it is clear that **SIMs can be considered as a comprehensive framework incorporating also many advanced models at a dynamic level.**

In this area also the **master equation/mean value equation models** (see Haag and Weidlich, 1984, and Haag, 1989b) have to be mentioned. These models have been used extensively in spatial flow analysis. This framework models the uncertainties in the decision process of the individuals via the master equation approach. The mean values are then obtained from the master equation by an aggregation of the individual probability distributions. Thus this approach provides the link between micro-economic aspects and the macro-economic equation of motions for aggregate mean values.

In this context also compartmental analysis should be mentioned (see De Palma and Lefevre, 1984) which consists of equations which are the approximate mean-value equations. It has also been shown (see Reiner et al., 1986) that these meanvalue equations may display chaotic behaviour with strange attractors, given particular conditions for the group interaction. On the other hand, this result is not surprising, since the mean value equations are strongly related to logit models. Hence it is plausible that interrelated logistic functions underlie the emerging chaotic motions.

After this brief review based on a typology of dynamic models and their potential in spatial interaction planning with respect to the scale of analysis, we will now show their evolution in comparison with the evolutionary pattern of SIMs.

9.3 Evolution of Spatial Interaction Models

The models discussed in the previous section may broadly be classified under the heading of DMs and then confronted, in their evolution, with the groups of static SIMs and DCMs. A further treatment of this issue, from the viewpoint of a life cycle interpretation, can be found in Nijkamp and Reggiani (1991c). As a synthesis, we can represent the series of the SIMs studied and adopted so far according to the scheme of overlapping generations illustrated in Figure 9.1.

In Figure 9.1 we have essentially depicted in an illustrative and indicative way the generation and diffusion of the three main families of models which have received a great deal of attention in the past century, i.e., SIMs, DCMs and DMs.

From Figure 9.1 we can see that - while SIMs present a smooth development at the beginning of the century with a much higher penetration after the sixties - DCMs and DMs exhibit a rapid growth (from both a theoretical and empirical point of view). It should be noted that we have included under the heading of DMs the whole group of dynamic models treated above, so that we can observe that after the mid seventies a broad variety of mathematical models has emerged.

Altogether, Figure 9.1 depicts overlapping generations of models; in particular we can unify all these models in an interesting general logistic shape evolving at present, where the points A and B mean the theoretical conjunction of DCMs and DMs with SIMs, respectively. However, we may plausibly expect that - given the logistic shape of Figure 9.1 - we are likely to approach a saturation level for the development of the above mentioned models. Besides, it is noteworthy that the sequential evolutionary paths of these models are not independent; they build upon each other's strength and make progress through synergetic effects.

Probably, from this upper saturation level onwards, new tools will emerge in the next decades, in response to new spatial activity patterns and new methodological progress. This upper level can likely be linked to the analytical structure of the models, since there are inevitable constraints in their formulation, so that also from a mathematical point of view the potential of such models is likely to reach a limit.

An example can be found by analyzing the contents and evolution of recent dynamic models (see Section 9.2). It has become clear that most of these models can be reformulated in terms of a logit-logistic formulation, and hence be interpreted in a SIM structure, so that the broad potential of SIMs can be considered to be such a limit in the form of an envelope.

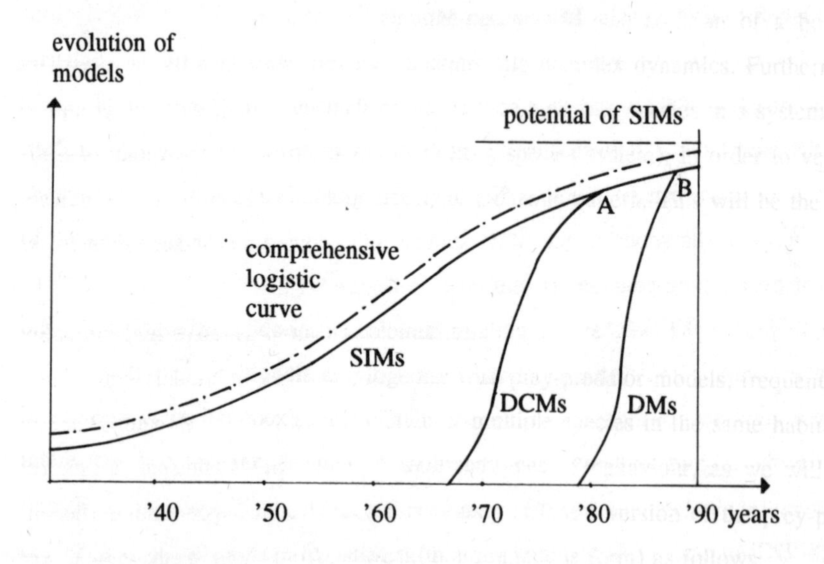

Figure 9.1 Evolution of Spatial Models
SIMs = Spatial Interaction Models
DCMs = Discrete Choice Models
DMs = Dynamic Models

9.4 New Research Areas

In the light of the above observations, it is clear that SIMs with their wealth of different formulations have governed the model scene in spatial analysis in the post-war period. Since a logistic shape can be connected with a dynamic SIM, it is plausible to assume that **many dynamic models of a different type** (such as master-equation models, prey-predator models, compartmental models, etc.) **can be derived and unified in the broad framework of SIMs**. The common background of such models makes them suitable for many applications in 'social physics' and in behavioural science.

In this context we can identify the following items of a research task which need further development:

- **empirical analysis**

Although currently a static analysis of SIMs can satisfactorily be carried out with most available data, many more difficulties emerge in the context of testing the empirical evidence of dynamic SIMs, owing to the lack of time series data. Consequently, some interesting properties such as chaotic movements, bifurcations, etc. are difficult to test empirically (see also Brock, 1986).

- **theoretical analysis**

From a **static** viewpoint the need for a further investigation of the relationships between SIMs and other choice methods both under conditions of certainty and uncertainty (such as multicriteria analysis, risk models, etc.) appeared to emerge in this study. Such an analysis may also generate other new approaches that are, besides the nested MNL model and the Alonso model, able to overcome the IIA axiom.

From a **dynamic** viewpoint it is necessary to develop closer connections between entropy, SIMs and chaos theory. In this context also a topological analysis of SIMs (for example, the study of a fractal character of spatial interaction phenomena) may offer new insights into the potential of SIMs.

Furthermore, in the context of a dynamic analysis, a new research direction will certainly be found in the study of the relationships between spatial interaction patterns in a complex network. In other words, given the increasing growth in information networks, transport flows and mobility patterns, it is important to pay attention to the analysis of the (in)stability of a network in relation to its degree of 'connectivity'. In this context new research developments have to be carried out in order to answer the following main questions:

- It is possible to specify behavioural hypotheses which lend themselves to empirical testing in a complex dynamic network configuration?
- Will an isolated stable subsystem remain stable by increasing the degree of connectivity of a network?
- If competition takes place between two 'prey' subsystems which usually generates instability in the whole system (or, vice versa, if competition between two 'predator' subsystems takes place which usually generates stability in the whole system), what is their impact - in terms of equilibrium tendencies - by increasing the connectivity of the network?
- Does dynamic systems theory provide us with new insights regarding the impact of time lags on the size and direction of flows in a non-linear dynamic network system?

and finally:

- Is it possible to apply new concepts from systems literature, such as autopoièsis (see Maturana and Varela, 1980), anagenesis (see, e.g., Boulding, 1978) or niche theory (see, e.g., Pianka, 1976) to non-biological systems such as networks of spatial interactions, communications and flows?

Consequently, it is clear that new analysis frameworks have to accompany, in the near future, the research in the area of SIMs, by taking into account new theories emerging from other disciplines, such as ecology, systems theory, biology, and artificial intelligence. By no means should behavioural research in the spatial sciences exclusively be based on a natural science background.

From the present study which was mainly devoted to the area of regional science, transportation science and geography, it appears that SIMs are a rich research area. In many cases, the evolution of SIMs seem to follow interesting evolutionary patterns characterized by stability and instability, equilibrium and disequilibrium, order and disorder. In any case, it is clear from the present study that SIMs can provide useful insight into the behaviour of complex spatial systems, even when up-to-date empirical information on such systems does not always exist, like in stochastic or chaotic processes. Especially in an applied setting however, much research remains to be done, on critical conditions for stability as well as on the collection of (longitudinal and panel) time series data.

References

Ahsan, S.M., The Treatment of Travel Time and Cost Variables in Disaggregate Mode Choice Models, **International Journal of Transport Economics**, vol. IX, 2, 1982, pp. 153-169.

Albin, P.S., Microeconomic Foundations of Cyclical Irregularities or Chaos, **Mathematical Social Sciences**, vol. 13, 1987, pp. 185-214.

Allen, P.M., J.L. Deneubourg, M. Sanglier, F. Boon and A. De Palma, The Dynamics of Urban Evolution, Final Report of the U.S. Department of Transportation, Washington D.C., 1978.

Alonso, W., A Theory of Movements, **Human Settlement Systems** (N.M. Hansen, ed.), Ballinger, Cambridge, Mass., 1978, pp. 195-212.

Anas, A., Discrete Choice Theory, Information Theory and the Multinomial Logit and Gravity Models, **Transportation Research B**, vol. 17, 1983, pp. 13-23.

Andersen, D.F. (ed.), **Chaos in System Dynamic Models: Special Issue System Dynamics Review**, vol. 4, no. 1/2, 1988, pp. 3-13.

Andersson, Å.E. and D.F. Batten, Creative Nodes, Logistical Networks and the Future of the Metropolis, **Transportation**, vol. 14, 1988, p. 281-293.

Anselin, L. and W. Isard, On Alonso's General Theory of Movement, **Man, Environment, Space and Time**, vol. 1, 1979, pp. 52-63.

Arnold, L., **Stochastic Differential Equations: Theory and Applications**, John Wiley, New York, 1974.

Aronson, D.G., M.A. Chory, G.R. Hall and R.P. McGehee, Bifurcations from an Invariant Circle for Two-Parameter Families of Maps of the Plane: A Computer Assisted Study, **Communications in Mathematical Physics**, vol. 83, 1982, pp. 303-354.

Arrow, K.J, **Social Choice and Individual Values**, John Wiley and Sons, New York, 1951.

Arrowsmith, G., A Behavioural Approach to Obtaining a Doubly Constrained Trip Distribution Model, **Operational Research Quarterly**, vol. 24, no. 1, 1973, pp. 100-111.

Baker, R.J. and J.A. Nelder, **The GLIM** (Generalized Linear Interactive Modelling) **System**, Numerical Algorithms Group, Oxford, 1978.

Balducci, R., G. Candela and G. Ricci, A Generalization of R. Goodwin's Model with Rational Behaviour of Economic Agents, **Non Linear Models of Fluctuating Growth** (Goodwin, R., M. Kruger and A. Vercelli, eds.), Lecture Notes in Economics and Mathematical Systems, vol. 228, Springer Verlag, Berlin, 1984, pp. 47-66.

Barentsen, W. and P. Nijkamp, Nonlinear Dynamic Modelling of Spatial Interactions, **Environment and Planning B**, 1988, vol. 15, pp. 433-446.

Barnett, W. and P. Chen, The Aggregation-Theoretic Monetary Aggregates Are Chaotic and Have Strange Attractors, **Dynamic Econometric Modelling** (Barnett, W., E. Berndt and H. White, eds.), Cambridge University Press, Cambridge, 1988, pp. 199-245.

Barras, R., T.A. Broadbent, M. Cordey-Hayes, D.B. Massey, K. Robinson and J. Willis, An Operational Urban Development Model of Cheshire, **Environment and Planning**, vol. 3, 1971, pp. 115-234.

Bartholomew, D.J., **Stochastic Models for Social Processes**, Wiley, London, 1973.

Batten, D.F., **Spatial Analysis of Interacting Economics**, Martinus Nijhoff, Dordrecht, The Netherlands, 1983.

Batten, D.F. and D.E. Boyce, Spatial Interaction, Transportation and Interregional Commodity Flow Models, **Handbook of Regional and Urban Economics** (P. Nijkamp, ed.), North-Holland, Amsterdam, 1986, pp. 357-406.

Batty, M., An Activity Allocation Model for the Nottinghamshire-Derbyshire Subregion, **Regional Studies**, vol. 4, 1970a, pp. 307-332.

Batty, M., Models and Projections of the Space Economy: A Subregional Study in Northwest England, **Town Planning Review**, vol. 41, 1970b, pp. 121-148.

Batty, M., Recent Developments in Land-Use Modelling: A Review of British Research, **Urban Studies**, vol. 9, 1972, pp. 151-177.

Batty, M., Spatial Entropy, **Geographical Analysis**, vol. 6, no. 1, 1974, pp. 1-32.

Batty, M., **Urban Modelling**, Cambridge University Press, Cambridge, 1976.

Batty, M., Cities as Fractals: Simulating Growth and Form, **Fractals and Chaos** (Crilly, A.J., R.A. Earnshaw and H. Jones, eds.), Springer-Verlag, Berlin, 1991, pp. 43-59.

Batty, M. and P.A. Longley, The Fractal Simulation of Urban Structure, **Environment and Planning A**, vol. 18, 1986, pp. 1143-1179.

Batty, M. and S. Mackie, The Calibration of Gravity, Entropy and Related Models of Spatial Interaction, **Environment and Planning**, vol. 4, no. 4, 1972, pp. 205-233.

Batty, M. and L. March, The Method of Residues in Urban Modelling, **Environment and Planning A**, vol. 8, no. 2, 1976, pp. 189-215.

Baumol, W.J. and J. Benhabib, Chaos: Significance, Mechanism, and Economic Applications, **Journal of Economic Perspectives**, vol. 3, no. 1, 1989, pp. 77-105.

Baumol, W.J. and E.N. Wolff, Feedback from Productivity Growth in R&D, **Scandinavian Journal of Economics**, vol. 85, 1983, pp. 145-157.

Beckmann, M.J., Entropy, Gravity and Utility in Transportation Modelling, **Information, Inference and Decision** (G. Menges, ed.), Riedel, Dordrecht, 1974, pp. 155-163.

Beckmann, M.J. and T.F. Golob, On the Metaphysical Foundations of Traffic Theory: Entropy Revisited, **Proceedings Fifth International Syposium on the Theory of Traffic Flow and Transportation** (G.F. Newell, ed.), Elsevier, New York, 1971.

Beckmann, M.J., C. McGuire and C.B. Winston, **Studies in the Economics of Transportation**, Yale University Press, New Haven, Connecticut, 1956.

Beckmann, M.J. and T. Puu, **Spatial Economics: Density, Potential and Flow**, North-Holland, Amsterdam, 1985.

Beckmann, M.J. and T. Puu, **Spatial Structures**, Springer Verlag, Berlin, 1990.

Ben-Akiva, M. and A. De Palma, Analysis of a Dynamic Residential Location Choice Model with Transaction Costs, **Journal of Regional Science**, vol. 26, 1986, pp. 321-341.

Ben-Akiva, M. and S.R. Lerman, Disaggregate Travel and Mobility Choice Models and Measures of Accessibility, **Behaviour Travel Modelling** (Hensher, D.A. and P.R. Stopher, eds.), Croom Helm, London, 1979, pp. 654-679.

Benhabib, J. and R.H. Day, Rational Choice and Erratic Behaviour, **Review of Economic Studies**, vol. 48, 1981, pp. 459-471.

Benhabib, J. and R.H. Day, A Characterization of Erratic Dynamics in the Overlapping Generations Model, **Journal of Economic Dynamics and Control**, vol. 4, 1982, pp. 37-55.

Berg, L. van den, L. Burns and L.H. Klaassen (eds.), **Spatial Cycles**, Gower, Aldershot, 1987.

Berry, B.J.L., Cities as Systems within Systems of Cities, **Papers and Proceedings of the Regional Science Association**, vol. 13, 1964, pp. 147-163.

Berry, B.J.L. and P. Schwind, Information and Entropy in Migrant Flows, **Geographical Analysis**, vol. 1, no. 1, 1969, pp. 5-14.

Bertuglia, C.S., G. Leonardi, S. Occelli, G.A. Rabino, R. Tadei and A.G. Wilson (eds.), **Urban Systems: Contemporary Approaches to Modelling**, Croom Helm, London, 1987.

Birkin, M. and M. Clarke, Comprehensive Dynamic Urban Models: Integrating Macro and Micro Approaches, **Evolving Geographical Structures** (Griffith, D.A. and A.C. Lea, eds.), 1983, pp. 164-191.

Birkin, M. and A.G. Wilson, Some Properties of Spatial-Structural-Economic-Dynamic Models, **Urban Dynamics and Spatial Choice Behaviour** (Hauer, J., H. Timmermans and N. Wrigley, eds.), Kluwer Academic Publishers, Dordrecht, 1989, pp. 175-201.

Boldrin, M., Persistent Oscillations and Chaos in Dynamic Economic Models: Notes for a Survey, **The Economy as an Evolving Complex System** (Arrow, K. and D. Pines, eds.), Santa Fe Institute Studies in the Science of Complexity, Addison-Wesley Publ. Co., Redwood City, CA, 1988, pp. 49-75.

Boldrin, M. and L. Montrucchio, On the Indeterminacy of Capital Accumulation Paths, **Journal of Economic Theory**, vol. 40, 1986, pp. 26-39.

Boldrin, M. and M. Woodford, Equilibrium Models Displaying Endogenous Fluctuations and Chaos: A Survey, Working Paper 530, UCLA Dept. of Economics, 1988.

Bonsall, P.W., P.J. Hills and M.Q. Dalvi (eds.), **Urban Transportation Planning - Current Themes and Future Prospects**, Abacus Press, Tunbridge Wells, 1977.

Bouchard, R.J. and C.E. Pyers, Use of Gravity Models for Describing Urban Travel, **Highway Research Record**, no. 88, 1964, pp. 1-43.

Boulding, K.E., **Ecodynamics: A New Theory of Social Evolution**, Sage, Beverly Hills, 1978.

Boyce D.E. (ed.), Transportation Research: the State of the Art and Research Opportunities, **Transportation Research A: Special Issue**, vol. 19, no. 5/6, 1985, pp. 349-542.

Boyce, D.E., K.S. Chon, Y.J. Lee, K.T. Lin and L.J. Le Blanc, Implementation and Computational Issues for Combined Models of Location, Destination, Mode and Route Choice, **Environment and Planning A**, vol. 15, 1983, pp. 1214-1230.

Brock, W.A., Distinguishing Random and Deterministic Systems: Abridged Version, **Journal of Economic Theory**, vol. 40, 1986, pp. 168-195

Brock, W.A., Chaos and Complexity in Economic and Financial Science, **Acting under Uncertainty: Multidisciplinary Conceptions** (G.M. von Furstenberg, ed.), Kluwer, Norwell, MA, 1989, pp. 101-123.

Brock, W.A. and C.L. Sayers, Is the Business Cycle Characterized by Deterministic Chaos? **Journal of Monetary Economics**, vol. 22, 1988, pp. 71-90.

Brock, W.A., W.D. Dechert and J.A. Scheinkman, A Test for Independence Based on the Correlation Dimension, SSRI Working Paper No. 8702, Department of Economics, University of Wisconsin at Madison, 1987.

Brouwer, F. and P. Nijkamp, Qualitative Structure Analysis of Complex Systems, **Measuring the Unmeasurable** (Nijkamp, P., H. Leitner and N. Wrigley, eds.), Martinus Nijhoff, Dordrecht, 1985, pp. 509-532.

Bunch, D.S., Estimatibility in the Multinomial Probit Model, Research Report UCD-TRG-RR-89-5, Transportation Research Group, University of California at Davis, Ca., 1989.

Bunch, D.S. and R. Kitamura, Multinomial Probit Model Estimation Revisited: Testing of New Algorithms and Evaluation of Alternative Model Specifications for Trinomial Models of Household Car Ownership, Research Report UCD-TRG-RR-89-4, Transportation Research Group, University of California at Davis, Ca., 1989.

Burnett, K.P., The Dimensions of Alternatives in Spatial Choice Processes, **Geographical Analysis**, vol. 5, 1973, pp. 181-204.

Bussiere, R. and F. Snickars, Derivation of the Negative Exponential Model by an Entropy Maximising Method, **Environment and Planning**, vol. 2, no. 3, 1970, pp. 295-301.

Campisi, D., Lotka-Volterra Models with Random Fluctuations for the Analysis of Oscillations in Urban and Metropolitan Areas, **Sistemi Urbani**, vol. 2/3, 1986, pp. 309-321.

Carey, H.C., **Principles of Social Science**, J.B. Lippincott, Philadelphia, PA, 1858.

Cartwright, T.J., Windows on Order and Chaos. The Role of Microcomputers in Planning and Management, Proceedings of the International Conference 'Computers in Urban Planning and Urban Management', Hong-Kong, August, 1989.

Casti, J.L., **Connectivity, Complexity and Catastrophe Theory in Large-Scale Systems**, John Wiley, New York, 1979.

Casti, J.L., **Alternate Realities: Mathematical Models of Nature and Man**, John Wiley, New York, 1989.

Cesario, F.J., A Note on the Entropy Model of Trip Distribution, **Transportation Research**, vol. 7, no. 3, 1973, pp. 331-333.

Cesario, F.J., Least Squares Estimation of Gravity Model Parameters, **Transportation Research**, vol. 9, 1975, pp. 13-18.

Cesario, F.J. and E.T. Zerdy, Urban Distribution Function, Spatial Entropy and Transformation Groups, Paper British Section of the Regional Science Association, London, 1975, (mimeographed).

Champernowne, A., H.C.W. L. Williams and J. Coelho, Some Comments on Urban Travel Demand Analysis, Model Calibration and the Economic Valuation of Transport Models, **Journal of Transport Economics and Policy**, vol. 10, 1976, pp. 267-285.

Charnes, A., W.R. Raike and O. Bettinger, An Extremal and Information-Theoretic Characterisation of some Interzonal Transfer Models, **Socio-Economic Planning Sciences**, vol. 6, no. 6, 1972.

Charnes, A., K.E. Haynes, F. Phillips and G.M. White, Solving the Gravity Model Using the Unconstrained Dual of an Extended Geometric Programming Problem, Austin, the University of Texas, 1976, (mimeographed).

Chauduri, K., A Control Theoretic Model of Multispecies Fish Harvesting, **Mathematical Modelling**, vol. 8, 1987, pp. 803-809.

Chiarella, C., The Cobweb Model: Its Instability and the Onset of Chaos, **Economic Modelling**, vol. 5, no. 4, 1988, pp. 377-384.

Chisholm, M. and P. O'Sullivan, **Freight Flows and Spatial Aspects of the British Economy**, Cambridge University Press, Cambridge, 1973.

Choukroun, J.M., A General Framework for the Development of Gravity-Type Trip Distribution Models, **Regional Science and Urban Economics**, vol. 5, no. 2, 1975, pp. 177-202.

Chow, G.C., Optimum Control of Stochastic Differential Equation Systems, **Journal of Economic Dynamics and Control**, vol. 1, 1979, pp. 143-175.

Clark, W.A.V., Structures for Research on the Dynamics of Residential Mobility, **Evolving Geographical Structures** (D.A. Griffith and A.C. Lea, eds.), Martinus Nijhoff, Dordrecht, 1983, pp. 372-397.

Clark, W.A.V. and M. Cadwallader, Residential Preferences: an Alternative View of Intra-Urban Space, **Environment and Planning**, vol. 5, 1973, pp. 693-703.

Clark, W.A.V. and T.R. Smith, Housing Market Search Behavior and Expected Utility Theory: 2. The Process of Search, **Environment and Planning A**, vol. 14, 1982, pp. 717-737.

Clarke, M. and A.G. Wilson, A Framework for Dynamic Comprehensive Urban Models: The Integration of Accounting and Microsimulation Approaches, **Sistemi Urbani**, no. 2/3, 1986, pp. 145-177.

Cochrane, R.A., A Possible Economic Basis for the Gravity Model, **Journal of Transport Economics and Policy**, vol. 9, no. 1, 1975, pp. 34-49.

Coelho, J.D., **The Use of Mathematical Methods in Model Based Land Use Planning**, Ph.D. Diss., School of Geography, University of Leeds, Leeds, 1977.

Coelho, J.D. and A.G. Wilson, Some Equivalence Theorems to Integrate Entropy Maximizing Sub-Models within Overall Mathematical Programming Frameworks, **Geographical Analysis**, vol. 9, 1977, pp. 160-173.

Coleman, J.S., **Longitudinal Data Analysis**, Basic Books, New York, 1981.

Collet, P. and J.P. Eckmann, **Iterated Maps on the Interval as Dynamical Systems**, Birkhauser, Boston, 1980.

Cordey-Hayes, H. and D. Gleave, Migration Movements and the Differential Growth of City Regions in England and Wales, **Papers of the Regional Science Association**, vol. 33, 1974, pp. 99-123.

Cordey-Hayes, M. and A.G. Wilson, Spatial Interaction, **Socio-Economic Planning Sciences**, vol. 5, no. 1, 1971, pp. 73-95.

Crecine, J.P., Tomm (Time Oriented Metropolitan Model), Technical Bulletin n. 6, Community Renewal Program, Pittsburgh, 1964.

Crilly, M.L., A Perspective of the Functions and Criticisms of Trip Distribution Models, **Operational Research Quarterly**, vol. 25, no. 1, 1974, pp. 111-123.

Crilly, T., The Roots of Chaos. A Brief Guide, **Fractals and Chaos** (Crilly, A.J., R.A. Earnshaw and H. Jones, eds.), Springer-Verlag, Berlin, 1991, pp. 193-209.

Crilly, A.J., R.A. Earnshaw and H. Jones (eds.), **Fractals and Chaos**, Springer-Verlag, Berlin, 1991.

Cripps, E.L. and E.A. Cater, The Empirical Development of a Disaggregated Residential Location Model: Some Preliminary Results, **Patterns and Processes in Urban and Regional Systems** (A.G. Wilson, ed.), Pion, London, 1971, pp. 114-145.

Crutchfield, J.P., J.D. Farmer and B.A. Huberman, Fluctuations and Simple Chaotic Dynamics, **Physics Reports**, vol. 92, 1982, pp. 45-82.

Crutchfield, J.P., J.D. Farmer, N.H. Packard and R.S. Shaw, **Scientific American**, vol. 255, no. 6, 1986, pp. 46-57.

Cugno, F. and L. Montrucchio, Teorema della Ragnatela. Aspettative Adattive e Dinamiche Caotiche, **Rivistà Internazionale di Scienze Economiche e Commerciali**, no. 8, 1984a, pp. 713-724.

Cugno, F. and L. Montrucchio, Some New Techniques for Modelling Non-Linear Economic Fluctuations, **Non-Linear Models of Fluctuating Growth** (Goodwin, R.M., M. Krüger and A. Vercelli, eds.), Springer-Verlag, Berlin, 1984b, pp. 146-165.

Curry, L., Division of Labour from Geographical Competition, **Annals of the Association of American Geography**, vol. 71, 1981, pp. 133-165.

Cushing, J.M. and M. Saleem, A Predator-Prey Model with Age Structure, **Journal of Mathematical Biology**, vol. 14, 1982, pp. 231-250.

Cvitanovic, P. (ed.), **Universality in Chaos**, Adam Hilger, Bristol, 1984.

Daganzo, C., **Multinomial Probit**, Academic Press, New York, 1979.

Dana, R.A. and L. Montrucchio, Dynamic Complexity in Duopoly Games, **Journal of Economic Theory**, vol. 40, 1986, pp. 40-56.

Day, R.H., Irregular Growth Cycles, **American Economic Review**, vol. 72, 1982, pp. 406-414.

De Grawe, P., and H. Dewachter, A Chaotic Monetary Model of the Exchange Rate, Discussion Paper no. 466, Centre for Economic Policy Research, London, 1990.

De La Barra, **Integrated Land Use and Transport Modelling**, Cambridge University Press, Cambridge, 1990.

Deming, W.E. and F.F. Stephan, On a Least Square Adjustment of Sampled Frequency Tables When Expected Marinal Totals Are Known, **Annals of Mathematical Statistics**, vol. 11, 1940, pp. 427-444.

Dendrinos, D.S., Catastrophe Theory in Urban and Transport Analysis, US Department of Transportation, Office of Systems Engineering, RSPA, Washington D.C., 1980.

Dendrinos, D.S., Turbulence and Fundamental Urban/Regional Dynamics, Paper presented at the American Association of Geographers, Washington D.C., April 1984a.

Dendrinos, D.S., The Decline of the U.S. Economy: A Perspective from Mathematical Ecology, **Environment and Planning A**, vol. 16, 1984b, pp. 651-662.

Dendrinos, D.S., On the Incongruous Spatial Employment Dynamics, **Technological Change, Employment and Spatial Dynamics** (P. Nijkamp, ed.), Springer-Verlag, Berlin, 1986, pp. 321-339.

Dendrinos, D.S., Volterra-Lotka Ecological Dynamics, Gravitational Interaction and Turbulent Transportation: An Integration, Paper presented at the 35th North American Meeting of the Regional Science Association, Toronto, Canada, 1988.

Dendrinos, D.S., **The Dynamics of Cities: Ecological Determinism, Dualism and Chaos**, Routledge, London, 1991 (forthcoming).

Dendrinos, D.S. and H. Mullally, Evolutionary Patterns of Urban Populations, **Geographical Analysis**, vol. 13, no. 4, 1981, pp. 328-344.

Dendrinos, D.S. and H. Mullally, Optimal Control in Non-Linear Ecological Dynamics of Metropolitan Areas, **Environment and Planning A**, vol. 15, 1983, pp. 543-550.

Dendrinos, D.S. and H. Mullaly, **Urban Evolution. Studies in the Mathematical Ecology of Cities**, Oxford University Press, Oxford, 1985.

Dendrinos, D.S. and M. Sonis, Variational Principles and Conservation Conditions in Volterra's Ecology and in Urban Relative Dynamics, **Journal of Regional Science**, vol. 26, no. 2, 1986, pp. 359-372.

Dendrinos, D.S. and M. Sonis, The Onset of Turbulence in Discrete Relative Multiple Spatial Dynamics, **Applied Mathematics and Computation**, vol. 22, 1987, pp. 25-44.

Dendrinos, D.S. and M. Sonis, Non Linear Discrete Relative Population Dynamics of the U.S. Regions, **Journal of Applied Mathematics and Computation**, vol. 25, 1988, pp. 265-285.

Dendrinos, D.S. and M. Sonis, **Turbulence and Socio-Spatial Dynamics**, Springer-Verlag, New York, 1990.

Deneckere, R. and S. Pelikan, Competitive Chaos, **Journal of Economic Theory**, vol. 40, 1986, pp. 13-25.

De Palma, A. and M. Ben-Akiva, An Interactive Dynamic Model of Residential Location Choice. Paper presented at the International Conference on Structural Economic Analysis and Planning in Time and Space, Umea, Sweden. Mimeographed, Center for Transportation Studies, Massachusetts Institute of Technology, Cambridge, 1981.

De Palma, A. and C. Lefèvre, Individual Decision Making in Dynamic Collective Systems, **Journal of Mathematical Sociology**, vol. 9, 1983, pp. 103-124.

De Palma, A. and C. Lefèvre, The Theory of Deterministic and Stochastic Compartmental Models and its Applications: the State of the Art, **Sistemi Urbani**, no. 3, 1984, pp. 281-323.

De Palma, A. and C. Lefèvre, Residential Change and Economic Choice Behaviour, **Regional Science and Urban Economics**, vol. 15, 1985, pp. 421-434.

De Palma, A. and C. Lefèvre, The Theory of Deterministic and Stochastic Compartmental Models and Its Applications, **Urban Systems: Contemporary Approaches to Modelling** (Bertuglia, C.S., G. Leonardi, S. Occelli, G.A. Rabino, R. Tadei and A.G. Wilson, eds.), Croom Helm, London, 1987, pp, 490-540.

Devaney, R.L., **Chaotical Dynamical Systems**, Benjamin-Cummings, Menlo Park, 1986.

Diappi, L. and A. Reggiani, Modelli di Trasporto e Geografia del Territorio, **Territorio e Trasporti. Modelli Matematici per l'Analisi e la Pianificazione** (A. Reggiani, ed.), Franco Angeli, Milano, 1985, pp. 125-147.

Dinkel, J.J., G.A. Kochenberger and S.N. Wong, Entropy Maximisation and Geometric Programming, **Environment and Planning A**, vol. 9, no. 4, 1977, pp. 419-427.

Domencich, T.A. and D. McFadden, **Urban Travel Demand: A Behavioural Analysis**, North Holland, Amsterdam, 1975.

Downs, R.M., Geographic Space Perception: Past Approaches and Future Prospects, **Progress in Geography**, vol. 2, 1970, pp. 75-108.

Duffin, R.J., E.L. Peterson and C.M. Zener, **Geometric Programming**, Wiley, New York, 1967.

Echenique, M. and J. Domeyko, A Model for Santiago Metropolitan Area, Land Use and Built Form Studies, University of Cambridge, 1970.

Echenique, M., D. Crowther and W. Lindsay, A Spatial Model of Urban Stock and Activity, **Regional Studies**, vol. 3, 1969, pp. 281-312.

Echenique, M., A. Feo, R. Herrera and J. Riquez, A Disaggregated Model of Urban Spatial Structure, **Environment and Planning A**, vol. 6, 1974, pp. 33-63.

Eckmann, J.P. and D. Ruelle, Ergodic Theory of Chaos and Strange Attractors, **Review of Modern Physics**, vol. 57, no. 3, 1985, pp. 617-656.

Erlander, S., Optimal Spatial Interaction and the Gravity Model, **Lecture Notes in Economics and Mathematical Systems**, vol. 173, Springer-Verlag, Berlin, 1980.

Erlander, S. and N.F. Stewart, **The Gravity Model in Transportation Analysis**, VSP, Utrecht, 1990.

Ermoliev, Y. and G. Leonardi, Some Proposals for Stochastic Facility Location Models, **Sistemi Urbani**, no. 3, 1981, pp. 455-470.

Est, J. van and J. van Setten, Least Squares Procedures for a Spatial Multiplicative Distribution Model, **New Developments in Modelling Travel Demand and Urban Systems** (Jansen, G.R.M., P.H.L. Bovy, J.P.J.M. van Est and F. Le Clercq, eds.), Saxon House, Farnborough, 1979, pp. 372-403.

Evans, A.W., The Calibration of Trip Distribution Models with Exponential or Similar Cost Functions, **Transportation Research**, vol. 5, no. 1, 1971, pp. 15-38.

Evans, S., A Relationship between the Gravity Model for Trip Distribution and the Transportation Problem in Linear Programming, **Transportation Research**, vol. 7, 1973, pp. 39-61.

Evans, S.P., Derivation and Analysis of Some Models for Combining Trip Distribution and Assignment, **Transportation Research**, vol. 10, 1976, pp. 37-57.

Evans, S.P. and H.R. Kirby, A Three-Dimensional Furness Procedure for Calibrating Gravity Models, **Transportation Research**, vol. 8, 1974, pp. 105-122.

Farmer, R.E.A., Deficit and Cycles, **Journal of Economic Theory**, vol. 40, 1986, pp. 77-88.

Fast, J.D., **Entropy**, MacMillan, London, 1970.

Feigenbaum, M.J., Quantitative Universality for a Class of Non-Linear Transformations, **Journal of Statistical Planning**, vol. 19, 1978, pp. 25-52.

Feller, W., **An Introduction to Probability Theory and its Applications**, Wiley, New York, 1968.

Finkenstädt, B. and P. Kuhbier, Chaotic Dynamics in Agricultural Markets, Free University of Berlin, 1990, mimeo.

Fisch, O., On the Utility of Entropy Maximization, **Geographical Analysis**, vol. 9, no. 1, 1977, pp. 79-84.

Fischer, M.M. and P. Nijkamp, Categorical Data and Choice Analysis in a Spatial Context, **Optimization and Discrete Choice in Urban Systems** (Hutchinson, B.G., P. Nijkamp and M. Batty, eds.), Springer Verlag, Berlin, 1985a, pp. 1-30.

Fischer, M.M. and P. Nijkamp, Developments in Explanatory Discrete Spatial Data and Choice Analysis, **Progress in Human Geography**, 1985b, vol. 9, pp. 515-551.

Fischer, M.M., P. Nijkamp and Y.Y. Papageorgiou, **Spatial Choices and Processes**, North-Holland, Amsterdam, 1990.

Florian M. and M. Gaudry, Transportation Systems Analysis: Illustrations and Extension of a Conceptual Framework, **Transportation Research B**, vol. 17, no. 2, 1983, pp. 147-153.

Florian M. and H. Spiess, The Convergence of Diagonalization Algorithms for Asymmetric Network Equilibrium Models, **Transportation Research B**, vol. 16, 1982, pp. 477-483

Flowerdew, R. and M. Aitken, A Method of Fitting the Gravity Model Based on the Poisson Distribution, **Journal of Regional Science**, vol. 22, 1982, pp. 191-202.

Foot, D., Spatial Interaction Modelling, **Models in Urban Geography** (C.S. Yadav, ed.), Concept Publ., New Delhi, 1986, pp. 51-69.

Fotheringham, A.S., A New Set of Spatial Interaction Models: The Theory of Competing Destinations, **Environment and Planning A**, vol. 15, 1983, pp. 15-36.

Fotheringham, A.S., Modelling Hierarchical Destination Choice, **Environment and Planning A**, vol. 18, 1986, pp. 401-418.

Fotheringham, A.S. and T. Dignan, Further Contributions to a General Theory of Movement, **Annals of the Association of American Geographers**, vol. 74, 1984, pp. 620-633.

Fotheringham, A.S. and M.E. O'Kelly, **Spatial Interaction Models: Formulations and Applications**, Kluwer Academic Publishers, Dordrecht, 1989.

Frankhauser, P., Aspects Fractals des Structures Urbaines, **L'Espace Géographique**, no. 1, 1991, pp. 45-69.

Friesz L.T., Transportation Network Equilibrium, Design and Aggregation: Key Developments and Research Opportunities, **Transportation Research A**, vol. 19, no. 5/6, 1985, pp. 413-427.

Funke, M., A Generalized Goodwin Model Incorporating Technical Progress and Variable Prices, **Economic Notes**, no. 2, 1987, pp. 36-47.

Furness, K.P., Time Function Iteration, **Traffic English Control**, vol. 7, 1965, pp. 458-460.

Gardiner, C.W., **Handbook of Stochastic Methods**, Springer-Verlag, Berlin, 1983.

Gaudry, M.J.I. and M.G. Dagenais, The Dogit Model, **Transportation Research B**, vol. 13, 1979, pp. 105-111.

Gause, G.F., **The Struggle for Existence**, Williams and Wilkins, Baltimore, 1934.

Georgescu-Roegen, N., **The Entropy Law and the Economic Process**, Harvard University Press, Cambridge, Mass., 1971.

Gilpin, M.E., Spiral Chaos in a Predator-Prey Model, **American Nature**, vol. 113, 1979, pp. 306-308.

Gilpin, M.E. and K.E. Justice, Reinterpretation of the Invalidation of the Principle of Competitive Exclusion, **Nature**, vol. 236, 1972, pp. 273-301.

Ginsberg, R.B., Critique of Probabilistic Models: Application of the Semi-Markov Model to Migration, **Journal of Mathematical Sociology**, vol. 2, 1972, pp. 63-82.

Goh, B.S., G. Leitmann and T. Vincent, Optimal Control of a Prey-Predator System, **Mathematical Biosciences**, vol. 19, 1974, pp. 263-286.

Goldner, W., Plum (Projective Land Use Model): A Model for the Spatial Allocation of Activities and Land Uses in a Metropolitan Region, BATSC Technical Report 219, Bay Area Transportation Study Commission, Berkeley, 1968.

Goldner, W. and R.S. Graybeal, Bass (The Bay Area Simulation Study): Pilot Model of Santa Clare County and Some Applications, Centre for Real Estate and Urban Economics, University of California, Berkeley, 1965.

Golledge, R.G. and L.A. Brown, Search, Learning and the Market Decision Process, **Geografiska Annaler**, vol. 49 B, 1967, pp. 116-124.

Golob, T.F. and M.J. Beckmann, A Utility Model for Travel Forecasting, **Transportation Science**, vol. 5, no. 1, 1971, pp. 79-90.

Golob, T.F., R.L. Gustafson and M.J. Beckmann, An Economic Utility Theory Approach to Spatial Interaction, **Papers of the Regional Science Association**, vol. 30, 1973, pp. 259-182.

Golob, T.F., H. Smith and L. van Wissen, Determination of Differences among Household Mobility Patterns, **Transportation and Mobility in an Era of Transition** (Jansen, G.R.M., P. Nijkamp and C. Ruijgrok, eds.), North-Holland, Amsterdam, 1984, pp. 211-226.

Good, I.J., Maximum Entropy for Hypothesis Formulation, Especially for Multidimensional Contigency Tables, **Annals of Mathematical Statistics**, vol. 34, 1963, pp. 911-934.

Goodwin, R., A Growth Cycle, **Socialism, Capitalism and Economic Growth**, Feinstein, Cambridge, 1967, pp. 54-58.

Gordon, I.R., Economic Explanations of Spatial Variation in Distance Deterrence, **Environment and Planning A**, vol. 17, 1985, pp. 59-72.

Gould, P., Pedagogic Review of A.G. Wilson's Entropy in Urban and Regional Modelling, **Annals of the Association of American Geographers**, vol. 62, no. 4, 1972, pp. 689-700.

Gould, P.R., On Mental Maps, Image and Environment, **Images of Spatial Environments** (Downs, R.M. and D. Stea, eds.), Aldine Press, Chicago, 1973, pp. 235-245.

Grandmont, J.M., On Endogenous Competitive Business Cycles, **Econometrica**, vol. 53, 1985, pp. 995-1046.

Grandmont, J.M. (ed.), **Special Issue Journal of Economic Theory**, vol. 40, 1986.

Gray, R.H. and A. Sen, Estimating Gravity Models Parameters: A Simplified Approach Based on the Odds Ratio, **Transportation Research B**, vol. 17, 1983, pp. 117-131.

Griffith, D.A. and A.C. Lea (eds.), **Evolving Geographical Structures**, Martinus Nijhoff, The Hague, 1983.

Guckenheimer, J. and P. Holmes, **Non-Linear Oscillations, Dynamical Systems and Bifurcation of Vector Fields**, Springer Verlag, Berlin, 1983.

Guckenheimer, J., G. Oster and A. Ipatchi, The Dynamics of the Density Dependent Population Models, **Journal of Mathematical Biology**, vol. 4, 1977, pp. 101-147.

Gurevitch, B.L., Geographical Differentiation and its Measures in a Discrete System, **Soviet Geography**, vol. 10, 1969, pp. 387-412.

Gylfason, T. (ed.), Papers and Proceedings of the Fifth Annual Congress of European Economic Association, **European Economic Review**, vol. 35, no.2/3, 1991.

Haag, G.A., Stochastic Theory for Residential and Labour Mobility, **Technological Change, Employment and Spatial Dynamics** (P. Nijkamp, ed.), Springer, Berlin, 1986, pp. 340-357.

Haag, G., **Dynamic Decision Theory: Applications to Urban and Regional Topics**, Kluwer, Dordrecht, 1989a.

Haag, G., Spatial Interaction Models and their Micro Foundation, **Advances in Spatial Theory and Dynamics**, (Andersson, Å.E., D.F. Batten, B. Johansson and P. Nijkamp, eds.), North-Holland, Amsterdam, 1989b, pp. 165-174.

Haag, G. and W. Weidlich, A Stochastic Theory of Interregional Migration, **Geographical Analysis**, vol. 16, 1984, pp. 331-357.

Haag, G. and W. Weidlich, A Dynamic Migration Theory and its Evaluation for Concrete Systems, **Regional Science and Urban Economics**, vol. 16, 1986, pp. 57-80.

Haken, H., **Synergetics**, Springer Verlag, Berlin, 1983a.

Haken, H., **Advanced Synergetics**, Springer Verlag, Berlin, 1983b.

Hallefjord, Å, and K. Jörnsten, Gravity Models with Multiple Objectives. Theory and Applications, **Transportation Research B**, vol. 20, 1986, pp. 19-39.

Halperin, W.C., The Analysis of Panel Data for Discrete Choices, **Measuring the Unmeasurable** (Nijkamp, P., H. Leitner and N. Wrigley, eds.), Martinus Nijhoff, Dordrecht, 1985, pp. 561-585.

Hannesson, R., Optimal Harvesting of Ecologically Interdependent Fish Species, **Journal of Environmental Economics and Management**, vol. 10, 1983, pp. 329-345.

Hansen, S., Utility, Accessibility and Entropy in Spatial Modelling, **Swedish Journal of Economics**, vol. 74, 1972, pp. 35-44.

Hansen, S., Entropy and Utility in Traffic Modelling, **Transportation and Traffic Theory** (D.T. Buckley, ed.), Elsevier, Amsterdam, 1975, pp. 435-452.

Hao, B.-L. (ed.), **Chaos**, World Scientific Publ. Co., Singapore, 1984.

Hao, B.-L., **Elementary Symbolic Dynamics and Chaos in Dissipative Systems**, World Scientific Publ. Co., Singapore, 1989.

Hardy, G.H., J.E. Littlewood and G. Polya, **Inequalities**, Cambridge University Press, Cambridge, Mass., 1967.

Harris, B. and A.G. Wilson, Equilibrium Values and Dynamics of Attractiveness Terms in Production-Constrained-Spatial-Interaction Models, **Environment and Planning A**, vol. 10, 1978, pp. 371-338.

Harris, B., J.- H. Chouckroun and A.G. Wilson, Economies of Scale and the Existence of Supply-Side Equilibria in a Production-Constrained Spatial-Interaction Model, **Environment and Planning A**, vol. 14, 1982, pp. 823-837.

Harsman, B. and F. Snickars, Disaggregated Housing Demand Models: Some Theoretical Approaches, **Papers of the Regional Science Association**, vol. 34, 1975, pp. 121-143.

Harvey, D., **Explanation in Geography**, Arnold, London, 1969.

Hastings, A., Age Dependent Predation is not a Simple Process, **Theoretical Population Biology**, vol. 23, 1983, pp. 347-362.

Hathaway, P.J., Trip Distribution and Disaggregation, **Environment and Planning A**, vol. 7, no. 1, 1975, pp. 71-97.

Hauer, J., H. Timmermans and N. Wrigley (eds.), **Urban Dynamics and Spatial Choice Behaviour**, Kluwer, Dordrecht, 1989.

Haurie, A. and G. Leitmann, On the Global Asymptotic Stability of Equilibrium Solutions for Open-Loop Differential Games, **Large Scale System Journal**, 1984, pp. 211-226.

Haynes, K.E. and A.S. Fotheringham, **Gravity and Spatial Interaction Models**, Sage Publ., Beverly Hills, 1984.

Heckman, J.J., Statistical Model for Discrete Panel Data, **Structural Analysis of Discrete Data with Econometric Applications** (Manski, C.F. and D. McFadden, eds.), MIT Press, Cambridge, Mass., 1981, pp. 114-178.

Heggie, I.G., Are Gravity and Interactance Models a Valid Technique for Planning Regional Transport Facilities?, **Operational Research Quarterly**, vol. 20, no. 1, 1969, pp. 93-110.

Hénon, M., A Two-Dimensional Mapping with a Strange Attractor, **Communications in Mathematical Physics**, vol. 50, 1976, pp. 69-77.

Hensher, D.A. and L.W. Johnson, **Applied Discrete Choice Modelling**, Croom Helm, London, 1981.

Hirsch, M. and S. Smale, **Differential Equations, Dynamical Systems and Linear Algebra**, Academic Press, London, 1974.

Hobson, A., **Concepts in Statistical Mechanics**, Gordon and Beach, New York, 1971.

Hobson, A. and B.K. Cheng, A Comparison of the Shannon and Kullback Information Measures, **Journal of Statistical Physics**, vol. 7, 1973, pp. 301-310.

Holden, A.V. (ed.), **Chaos**, Manchester University Press, Manchester, 1986.

Holden, A.V. and M.A. Muhamad, A Graphical Zoo of Strange and Peculiar Attractors, **Chaos** (A.V. Holden, ed.), Manchester University Press, Manchester, 1986, pp. 15-35.

Holden, D.J., Wardrop's Third Principle, **Journal of Transport Economics and Policy** vol. XXIII, no. 3, 1989, pp. 239-262.

Hommes, C.H. and H.E. Nusse, Does an Unstable Keynesian Unemployment Equilibrium in a Non-Walrasian Dynamic Macroeconomic Model Imply Chaos?, **Scandinavian Journal of Economics**, vol. 91, 1989, pp. 161-167.

Hommes, C.H., H.E. Nusse and A. Simonovits, Hicksian Cycles and Chaos in a Socialist Economy, Research Memorandum, Institute for Economic Research, State University, Groningen, 1990.

Horowitz, J., Travel Demand Analysis. State of the Art and Research Opportunities, **Transportation Research**, vol. 19, 1985, pp. 441-453.

Hua, C., An Exploration of the Nature and Rationale of a Systemic Model, **Environment and Planning A**, vol. 12, 1980, pp. 713-726.

Huff, D., Defining and Estimating a Trading Area, **Journal of Marketing**, 1964, vol. 28, pp. 34-38.

Hutchinson, G.E., Circular Causal Systems in Ecology, **Annals of the New York Academy of Science**, vol. 50, 1948, pp. 221-246.

Hyman, G.M., The Calibration of Trip Distribution Models, **Environment and Planning A**, vol. 1, 1969, pp. 105-112.

Ingels, F.M., **Information and Coding Theory**, Intext, Scranton, 1971.

Iooss, G., **Bifurcations of Maps and Application**, North-Holland, New York, 1979.

Isard, W., **Methods of Regional Analysis**, M.I.T. Press, Cambridge, Mass., 1960.

Isard, W. and P. Liossatos, **Spatial Dynamics and Optimal Space-Time Development**, North-Holland, Amsterdam, 1979.

Isard, W. and V.W. Maclaren, Storia e Stato Attuale delle Ricerche nella Scienza Regionale, **Problematiche dei Livelli Subregionali di Programmazione** (Bielli, M. and A. La Bella, eds.), Franco Angeli, Milano, 1982, pp. 19-35.

Jansen, G.R.M, P.H.L. Bovy, J.P.J.M. van Est and F. le Clercq (eds.), **New Development in Modelling Travel Demand and Urban Systems**, Saxon House, Farnborough, 1979, pp. 281-319.

Jaynes, E.T., Information Theory and Statistical Mechanics, **Physical Review**, vol. 106, 1957, pp. 620-630.

Jaynes, E.T., Prior Probabilities, **IEEE Transactions on Systems Science and Cybernetics**, SSC-4, 1968, pp. 227-241.

Jaynes, E.T., The Well-Posed Problem, **Foundations of Physics**, vol. 3, 1973, pp. 477-492.

Johansson, B. and P. Nijkamp, Analysis of Episodes in Urban Event Histories, **Spatial Cycles** (van den Berg, L., L.S. Burns and L.H. Klaassen, eds.), Gower, Aldershot, 1987, pp. 43-66.

Kaashoek, J.F. and A.C.F. Vorst, The Cusp Catastrophe in the Urban Retail Model, **Environment and Planning A**, vol. 16, 1984, pp. 851-862.

Kadás, S.A. and E. Klafszky, Estimation of the Parameters in the Gravity Model for Trip Distribution, **Regional Science and Urban Economics**, vol. 6, no. 4, 1976, pp. 439-457.

Kahn, D., J.L. Deneubourg and A. de Palma, Transportation Mode Choice, **Environment and Planning A**, vol. 13, 1981, pp. 1163-1174.

Kahn, D., J.L. Deneubourg and A. de Palma, Transportation Mode Choice and City-Suburban Public Transportation Service, **Transportation Research B**, vol. 17, 1982, pp. 25-43.

Kamien, M.I. and N.L. Schwartz, **Dynamic Optimization: The Calculus of Variations and Optimal Control in Economics and Managements**, North-Holland, Amsterdam, 1981.

Kanaroglou, P., K.L. Liaw and Y.Y Papageorgiou, An Analysis of Migratory Systems: 1. Theory, **Environment and Planning A**, vol. 18, 1986a, pp. 913-928.

Kanaroglou, P., K.L. Liaw and Y.Y. Papageorgiou, An Analysis of Migratory Systems: 2. Operational Framework, **Environment and Planning A**, vol. 18, 1986b, pp. 1039-1060.

Kaplan, W., **Ordinary Differential Equations**, Addison-Wesley Publ. Co., Reading, 1958.

Kelsey, D., The Economics of Chaos or the Chaos of the Economics, **Oxford Economic Papers**, vol. 40, 1988, pp. 1-31.

Kirby, H.R., Normalising Factors of the Gravity Model - An Interpretation, **Transportation Research**, vol. 4, 1970, pp. 37-50.

Kirby, H.R., Theoretical Requirements for Calibrating Gravity Models, **Transportation Research**, vol. 8, 1974, pp. 97-104.

Koçak, H., **Differential and Difference Equations through Computer Experiments**, Springer, Berlin, 1986.

Koppelman, F.S. and E.I. Pas, Travel-Activity Behavior in Time and Space, **Measuring the Unmeasurable** (Nijkamp, P., H. Leitner and N. Wrigley, eds.), Martinus Nijhoff, Dordrecht, 1985, pp. 587-627.

Ku, Y.H., **Analysis and Control of Non Linear Systems**, The Ronald Press Company, New York, 1958.

Kullback, S., **Statistics and Information**, Wiley, New York, 1959.

Kurths, D.A. and H. Herzel, An Attractor in Solar Time Series, **Physica D**, vol. 25, 1987, pp. 165-172.

Kushner, H., **Introduction to Stochastic Control**, Holt, Reineheart and Winston, New York, 1971.

Lakshmanan, T.R. and W.G. Hansen, A Retail Market Potential Model, **Journal of the American Institute of Planners**, vol. 31, 1965, pp. 134-143.

Lasota, A. and M.C. Mackey, **Probabilistic Properties of Deterministic Systems**, Cambridge University Press, Cambridge, 1985.

Lauwerier, H.A., Two-Dimensional Iterative Maps, **Chaos** (A.V. Holden, ed.), Manchester University Press, Manchester, 1986, pp. 58-95.

Ledent, J., Calibrating Alonso's General Theory of Movement: The Case of Interprovincial Migration Flows in Canada, **Sistemi Urbani**, vol. 2, 1980, pp. 327-358.

Ledent, J., On the Relationship between Alonso's Theory of Movement and Wilson's Family of Spatial Interaction Models, **Environment and Planning A**, vol. 13, 1981, pp. 217-224.

Leonardi, G., A Unifying Framework for Public Facility Location Problems - Part I: A Critical Review and Some Unsolved Problems, **Environment and Planning A**, vol. 13, 1981a, pp. 1085-1108.

Leonardi, G., A Unifying Framework for Public Facility Location Problems - Part II: Some New Models and Extensions, **Environment and Planning A**, vol. 13, 1981b, pp. 1085-1108.

Leonardi, G., A General Accessibility and Congestion-Sensitive Multiactivity Spatial Interaction Model, **Papers of the Regional Science Association**, vol. 47, 1981c, pp. 3-16.

Leonardi, G., An Optimal Control Representation of Stochastic Multistage Multiactor Choice Process, **Evolving Geographical Structure** (Griffith, D.A. and A.C. Lea, eds.), Martinus Nijhoff, The Hague, 1983, pp. 62-72.

Leonardi, G., Equivalenza Asintotica tra la Teoria delle Utilità Casuali e la Massimizzazione dell'Entropia, **Territorio e Trasporti** (A. Reggiani, ed.), Franco Angeli, Milano, 1985a, pp. 29-66.

Leonardi, G., A Stochastic Multistage Mobility Choice Model, **Optimization and Discrete Choice in Urban Systems** (Hutchinson, G., P. Nijkamp and H. Batty, eds.), Springer Verlag, Berlin, 1985b, pp. 132-147.

Leonardi, G. and D. Çampisi, Dynamic Multistage Random Utility Choice Processes: Models in Discrete and Continuous Time, Paper presented at the 2nd Meeting of the Italian Section of the Regional Science Association, Rome, 1981.

Li, T.Y. and J.A. Yorke, Period Three Implies Chaos, **American Mathematical Monthly**, vol. 82, no. 10, 1975, pp. 985-992.

Liaw, K.L. and J. Schuur, Application of a Generalized Nested Logit Model to the Explanation of Interprovincial Migration in the Netherlands: An Analysis Based on Housing Survey Data, Working Paper no. 76, Netherlands Interuniversity Demographic Institute, The Hague, 1988.

Lichtenberg, A.J. and M.A. Lieberman, **Regular and Stochastic Motion**, Springer Verlag, Berlin, 1983.

Lierop, W.F.J. van and P. Nijkamp, Spatial Choice and Interaction Models; Criteria and Aggregation, **Urban Studies**, vol. 17, 1980, pp. 229-311.

Lierop, W.F.J. van and P. Nijkamp, Disaggregate Models of Choice in a Spatial Context, **Sistemi Urbani**, Vol. 4, 1982, pp. 331-369.

Lierop, W.F.J. van and P. Nijkamp, Disaggregate Residential Choice Models; A Case Study for the Netherlands, **Scandinavian Housing and Planning Research**, no. 8, 1991, pp. 133-151.

Lierop, W.F.J. van and A. Rima, Trends and Prospects for Qualitative Disaggregate Spatial Choice Models, **Measuring the Unmeasurable** (Nijkamp, P., H. Leitner and N. Wrigley, eds.), Martinus Nijhoff, Dordrecht, 1985, pp. 533-560.

Lindley, D.V., **Bayesian Statistics: A Review**, Society for Industrial and Applied Mathematics, Philadelphia, 1972.

Lombardo, S.T. and G.A. Rabino, Some Simulations of a Central Place Theory Model, **Sistemi Urbani**, vol. 5, 1983, pp. 315-332.

Longley, P., Discrete Choice Modelling and Complex Spatial Choice: An Overview, **Recent Developments in Spatial Data Analysis** (Bahrenberg, G., M.M. Fischer and P. Nijkamp, eds.), Gower, Aldershot, Hants, 1984, pp. 375-391.

Lorenz, E.N., Deterministic Non-Periodic Flow, **Journal of the Atmospheric Sciences**, vol. 20, no. 2, 1963, pp. 130-141.

Lorenz, H.-W., International Trade and the Possible Occurrence of Chaos, **Economic Letters**, 1987, vol. 23, pp. 135-138.

Lorenz, H.-W., **Non Linear Dynamical Economics and Chaotic Motion**, Lectures Notes in Economics and Mathematical Systems, vol. 334, Springer Verlag, Berlin, 1989.

Lotka, A.J., **Elements of Physical Biology**, Williams and Wilkins, Baltimore, 1925.

Lowe, J.C. and S. Moryades, **The Geography of Movement**, Houghton Mifflin Co., Boston, 1975.

Lowry, I.S., A Model of Metropolis, RM-4035-RC, The Rand Corporation, Santa Monica, 1964.

Luce, R.D., **Individual Choice Behaviour: A Theoretical Analysis**, John Wiley and Sons, New York, 1959.

Luce, R.D. and H. Raiffa, **Games and Decision**, John Wiley and Sons, New York, 1957.

Lucking, R., Chaos. The Origins and Relevance of a New Discipline, **Project Appraisal**, vol. 6, no. 1, 1991, pp. 1-64.

Lung, Y., Complexity and Spatial Dynamics Modelling, **The Annals of Regional Science**, vol. 22, no. 2, 1988, pp. 81-111.

MacGill, S.M., Theoretical Properties of Biproportional Matrix Adjustments, **Environment and Planning A**, vol. 9, 1977, pp. 687-701.

MacGill, S.M., Convergence and Related Properties for a Modified Biproportional Matrix Problem, **Environment and Planning A**, vol. 11, 1979, pp. 499-506.

Malliaris, A.G. and W.A. Brock, **Stochastic Methods in Economics and Finance**, North-Holland, Amsterdam, 1982.

Mandelbrot, B., **The Fractal Geometry of Nature**, W.H. Freeman and Company, New York, 1977.

Manheim, M.L., **Fundamentals of Transportation Systems Analysis**, vol. 1: Basic Concepts, MIT Press, Cambridge, Mass., 1979.

Manneville, P. and Y. Pomeau, Intermittency and the Lorenz Model, **Physics Letters A**, vol. 75, no. 1/2, 1979, pp. 1-2.

Manski, C.F. and D. McFadden, **Structural Analysis of Discrete Data with Econometric Applications**, MIT Press, Cambridge, Mass., 1981.

Marsden, J.E. and M. McCracken, **The Hopf Bifurcation and its Applications**, Springer Verlag, Berlin, 1976.

Masser, I. and P. Brown (eds.), **Spatial Representation and Spatial Interaction**, Martinus Nijhoff, The Hague, 1977.

Mathai, A. and P. Rathie, **Basic Concepts in Information Theory and Statistics**, Halsted/Wiley, New York, 1975.

Matis, J.H. and Hartley, H.O., Stochastic Comparmental Analysis: Model and Least Squares Estimation from Time Series Data, **Biometrics**, vol. 35, 1971, pp. 77-102.

Mattsson, L.G., Equivalence between Welfare and Entropy Approaches to Residential Location, Research Paper TRITA-MAT-1983-18, Royal Institute of Technology, Stockholm, 1983.

Maturana, H.M. and F.G. Varela (eds.), **Autopoiesis and Cognition: The Realization of the Living**, Reidel, Dordrecht, 1980.

May, R., Biological Populations with Nonoverlapping Generations, Stable Points, Stable Cycles and Chaos, **Science**, vol. 186, 1974, pp. 645-647.

May, R., Simple Mathematical Models with Very Complicated Dynamics, **Nature**, vol. 261, 1976, pp. 459-467.

Maynard Smith, J., **Mathematical Ideas in Biology**, Cambridge University Press, Cambridge, 1968.

Maynard Smith, J., **Models in Ecology**, Cambridge University Press, London, 1974.

McDonald, J.F., **Economic Analysis of an Urban Housing Market**, Academic Press, New York, 1979.

McFadden, D., Conditional Logit Analysis of Qualitative Choice Behaviour, **Frontiers in Econometrics** (P. Zarembka, ed.), Academic Press, New York, 1974, pp. 105-142.

McFadden, D., Modelling the Choice of Residential Location, **Spatial Interaction Theory and Planning Models** (Karlqvist, A., L. Lundqvist, F. Snickars, and J.W. Weibull, eds.), North-Holland, Amsterdam, 1978, pp. 75-96.

McFadden, D., Qualitative Methods for Analysing Travel Behaviour of Individuals: Some Recent Developments, **Behaviour Travel Modelling** (Hensher, D.A. and P.R. Stopher, eds.), Croom Helm, London, 1979, p. 279-318.

Medvedkov, Y.V., The Concept of Entropy in Settlement Pattern Analysis, **Papers of the Regional Science Association**, vol. 18, 1967, pp. 165-168.

Mees, A., The Revival of Cities in Medieval Europe: An Application of Catastrophe Theory, **Regional Science and Urban Economics**, vol. 5, 1975, pp. 403-425.

Merkies, A.H.Q.M. and T. van der Meer, Scope of the Three-Component Model, **Regional Science and Urban Economics**, vol. 19, 1989, pp. 601-614.

Miller, R.E., **Dynamic Optimization and Economic Applications**, McGraw-Hill, New York, 1979.

Miyao, T. and P. Shapiro, Discrete Choice and Variable Returns to Scale, **International Economic Review**, vol. 22, no. 2, 1981, pp. 257-273.

Mosekilde, E., J. Aracil and P.M. Allen, Instabilities and Chaos in Non-Linear Dynamic Systems, **System Dynamics Review**, vol. 4, no.1/2, 1988, pp. 14-55.

Newhouse, S., D. Ruelle and F. Takens, Occurrence of Strange Axiom - A Attractors near Quasiperiodic Flow on T^m, $m \geq 3$, **Communications in Mathematical Physics**, vol. 64, 1978, pp. 35-40.

Nicolis, G. and I. Prigogine, **Self-Organization in Non-Equilibrium Systems**, Wiley, New York, 1977.

Niedercorn, J.H. and B.V. Bechdolt, An Economic Derivation of the 'Gravity Law' of Spatial Interaction, **Journal of Regional Science**, vol. 9, 1969, pp. 273-281.

Nijkamp, P., **Planning of Industrial Complexes by Means of Geometric Programming**, Rotterdam University Press, Rotterdam, 1972.

Nijkamp, P., Reflections on Gravity and Entropy Models, **Regional Science and Urban Economics**, vol. 5, 1975, pp. 203-225.

Nijkamp, P., Spatial Mobility and Settlement Patterns: An Application of a Behavioural Entropy, International Institute for Applied Systems Analysis, Laxenburg, 1976, Research Memorandum RM 76-45.

Nijkamp, P., **Theory and Application of Environmental Economics**, North-Holland, Amsterdam, 1977.

Nijkamp, P., A Theory of Displaced Ideals: an Analysis of Interdependent Decisions via Nonlinear Multiobjective Optimization, **Environment and Planning A**, vol. 11, 1979, pp. 1165-1178.

Nijkamp, P., **Environmental Policy Analysis: Operational Methods and Models**, John Wiley & Sons, Chichester, New York, 1980.

Nijkamp, P., Long Term Economic Fluctuations: A Spatial View, **Socio-Economic Planning Sciences**, vol. 21, no. 3, 1987a, pp. 189-197.

Nijkamp, P., Discrete Spatial Choice Analysis: Special Issue, **Regional Science and Urban Economics**, vol. 17, no. 1, 1987b, pp. 1-27.

Nijkamp, P., Regularity and Chaos in Economics: Are Economic Predictions Useless?, Research Memorandum, Department of Economics, Free University, Amsterdam, 1990.

Nijkamp, P. and J.H.P. Paelinck, A Dual Interpretation and Generalisation of Entropy-Maximisation Models in Regional Science, **Papers of the Regional Science Association**, vol. 33, 1974, pp. 14-31.

Nijkamp, P. and J.H.P. Paelinck, **Operational Theory and Method of Regional Economics**, Saxon House, Farnborough, 1976.

Nijkamp, P. and J. Poot, Dynamics of Generalized Spatial Interaction Models, **Regional Science and Urban Economics**, vol. 17, 1987, pp. 367-390.

Nijkamp, P. and J. Poot, Lessons from Non-Linear Dynamic Economics, Research Memorandum, Department of Economics, Free University, Amsterdam, 1991.

Nijkamp, P. and A. Reggiani, Entropy, Spatial Interaction Models and Discrete Choice Analysis: Static and Dynamic Analogies, **European Journal of Operational Research**, vol. 36, no. 1, 1988a, pp. 186-196.

Nijkamp, P. and A. Reggiani, Analysis of Dynamic Spatial Interaction Models by Means of Optimal Control, **Geographical Analysis**, vol. 20, no. 1, 1988b, pp. 18-29.

Nijkamp, P. and A. Reggiani, Dynamic Spatial Interaction Models: New Directions, **Environment and Planning A,** vol. 20, 1988c, pp. 1449-1460.

Nijkamp, P. and A. Reggiani, Spatial Interaction and Discrete Choice: Statics and Dynamics, **Urban Dynamics and Spatial Choice Behaviour** (Hauer, J., H. Timmermans and N. Wrigley, eds.), Kluwer Academic Publishers, Dordrecht, 1989a, pp. 125-151.

Nijkamp, P. and A. Reggiani, Spatial Interaction and Input-Output Models: A Dynamic Stochastic Multi-objective Framework, **Frontiers of Input-Output Analysis** (Miller, R., K. Polenske and A. Rose, eds.), Oxford Press, London, 1989b, pp. 193-205.

Nijkamp, P. and A. Reggiani, An Evolutionary Approach to the Analysis of Dynamic Systems with Special Reference to Spatial Interaction Models, **Sistemi Urbani**, vol. 1, 1990a, pp. 601-614.

Nijkamp, P. and A. Reggiani, Logit Models and Chaotic Behaviour: A New Perspective, **Environment and Planning A,** vol. 22, 1990b, pp. 1455-1467.

Nijkamp, P. and A. Reggiani, Theory of Chaos: Relevance for Analysing Spatial Processes, **Spatial Choices and Processes** (Fischer, M.M., P. Nijkamp and Y.Y. Papageorgiou, eds.), North-Holland, Amsterdam, 1990c, pp. 49-79.

Nijkamp, P. and A. Reggiani, Spatio-Temporal Processes in Dynamic Logit Models, **Occasional Paper Series on Socio-Spatial Dynamics**, vol. 1, no. 1, 1990d, pp. 21-40.

Nijkamp, P. and A. Reggiani, Chaos Theory and Spatial Dynamics, **Journal of Transport Economics and Policy,** vol. XXV, no. 1, 1991a, pp. 81-96.

Nijkamp, P. and A. Reggiani, Space Time Dynamics, Spatial Competition and the Theory of Chaos, **Proceedings Annual Conference Italian Regional Science Association**, Taormina, 1991b, vol. 1, pp. 211-238.

Nijkamp, P. and A. Reggiani, Strategic Long-Term Transportation Planning and Dynamic Transportation Models: A Life Cycle Interpretation, **Sistemi Urbani**, vol. 1/2, 1991c.

Nijkamp, P. and A. Reggiani, Impacts of Multiple Period Lags in Dynamic Logit Models, **Geographical Analysis**, vol. 24, no. 2, 1992.

Nijkamp, P. and S. Reichman, **Transportation Planning in a Changing World**, Gower, Aldershot, UK, 1988.

Nijkamp, P, and P. Rietveld, Multiple Objective Decision Analysis in Regional Economics, **Handbook of Regional and Urban Economics** (Nijkamp, P., ed.), North-Holland, Amsterdam, 1986, pp. 493-541.

Nijkamp, P., H. Leitner and N. Wrigley (eds.), **Measuring the Unmeasurable: Analysis of Qualitative Spatial Data**, Martinus Nijhoff, Dordrecht, 1985.

Nijkamp, P., J. Poot and J. Rouwendal, A Non-Linear Dynamic Model of Spatial Economic Development and R&D Policy, **Annals of Regional Science**, vol. 25, no. 4, 1991, pp. 287-302.

Nusse, H.E. and C.H. Hommes, Resolution of Chaos with Application to a Modified Samuelson Model, **Journal of Economic Dynamics and Control**, vol. 14, 1990, pp. 1-19.

Odum, E.P., **Fundamentals of Ecology**, Saunders, Philadelphia, Penn., 1971.

Oertel, H., Nonlinear Dynamics. Temporal and Spatial Structures in Fluid Mechanics, **Nonlinear Dynamics of Transcritical Flows** (Jordan, H.L., H. Oertel and K. Robert), **Lecture Notes in Engineering**, Springer Verlag, Berlin, 1984, pp. 1-36.

Okabe, A., Population Dynamics of Cities in a Region: Conditions for a State of a Simultaneous Growth, **Environment and Planning A**, vol. 11, 1979, pp. 609-628.

Openshaw, S., An Empirical Study of Some Spatial Interaction Models, **Environment and Planning A**, vol. 8, no. 1, 1976, pp. 23-41.

Openshaw, S., Optimal Zoning Systems for Spatial Interaction Models, **Environment and Planning A**, vol. 9, 1977, pp. 169-184.

Oppenheim, N., Dynamic Forecasting of Urban Shopping Travel Demand, **Transportation Research B**, vol. 20, no. 5, 1986, pp. 391-402.

Orishimo, I., An Approach to Urban Dynamics, **Geographical Analysis**, vol. 19, no. 3, 1987, pp. 200-210.

Ornstein, D.S. and B. Weiss, Statistical Properties of Chaotic Systems, **Bulletin of the American Mathematical Society**, vol. 24, no. 1, 1991, pp. 11-116.

Ott, E., Strange Attractors and Chaotic Motions of Dynamical Systems, **Review of Modern Physics**, vol. 53, no. 4, 1981, pp. 655-671.

Pacini, P.M., Recent Approaches in Dynamical System Theory in Economics with Emphasis on Chaos: A Critic Review, European University Institute, Florence, 1986 (mimeographed).

Peitgen, H.O. and P.H. Richter, **The Beauty of Fractals**, Springer Verlag, Berlin, 1986.

Peters, T., **Thriving on Chaos**, MacMillan, London, 1988.

Phiri, P., Calculation of the Equilibrium Configuration of Shopping Facility Sizes, **Environment and Planning A**, vol. 12, 1980, pp. 983-1000.

Pianka, E.R., **Evolutionary Ecology**, Harper & Row, Publ., New York, 1978.

Pickles, A., Models of Movement: A Review of Alternative Methods, **Environment and Planning A**, vol. 12, 1980, pp. 1383-1404.

Pluym, W.K., The Application of Information - Theory to Transportation Research: A Systematic Framework, **New Developments in Modelling Travel Demand and Urban Systems** (Jansen, G.R.M., P.H.L. Bovy, J.P.J.M. van Est and F. Le Clercq, eds.), Saxon House, Farnborough, 1979, pp. 320-343.

Pohjola, M.T., Stable and Chaotic Growth: The Dynamics of a Discrete Version of Goodwin's Growth Cycle Model, **Zeitschrift für Nationalökonomie**, vol. 41, 1981, pp. 27-38.

Poincaré, H., **Les Méthods Nouvelles de la Méchanique Celeste**, Gauthier-Villers, Paris, 1892.

Poston, J.M.T. and H.B. Stewart, **Non-Linear Dynamics and Chaos**, John Wiley & Sons, Chichester, 1986.

Pounder, J.R. and T.D. Rogers, The Geometry of Chaos: Dynamics of a Non-Linear Second-Order Difference Equation, **Bulletin of Mathematical Biology**, vol. 42, 1980, pp. 551-597.

Prigogine, I. and I. Stengers, **Order out of Chaos**, Fontana, London, 1985.

Prskawetz, A., Are Demographic Indicators Deterministic or Stochastic?. Institute for OR, Vienna Technical University, 1990, mimeo.

Putman, S.H., Dram (Calibrating a Disaggregated Residential Allocation Model, **Alternative Frameworks for Analysis** (Massey, D.B. and P.W.J. Batey, eds.), London Papers in Regional Science, vol. 7, Pion, London, 1977.

Puu, T., **Non-Linear Economic Dynamics**, Springer Verlag, Berlin, 1989.

Quigley, J.M., What Have We Learned about Housing Markets?, **Recent Issues in Urban Economics** (Mieszkowski, P. and M.R. Straszheim, eds.), Baltimore, The John Hopkins University Press, 1979, pp. 391-429.

Rabino, G.A. and S.T. Lombardo, A Compartmental Analysis of Residential Mobility in Turin, **Sistemi Urbani**, vol. 2/3, 1986, pp. 263-284.

Radzicki, M.J., Institutional Dynamics, Deterministic Chaos and Self-Organizing Systems, **Journal of Economic Issues**, vol. 24, 1990, pp. 57-102.

Ragozin, D.L. and G. Brown, Harvest Policies and Non Market Valuation in a Predator-Prey System, **Journal of Environmental Economics and Management**, vol. 12, 1985, pp. 155-168.

Ramsey, J.P., C.L. Sayers and P. Rothman, The Statistical Properties of Dimension Calculations Using Small Data Sets: Some Economic Applications, **International Economic Review**, vol. 31, 1990, pp. 991-1020.

Rand, D., Exotic Phenomena in Games and Duopoly Models, **Journal of Mathematical Economics**, vol. 5, 1978, pp. 173-184.

Rasmussen, D.R. and E. Mosekilde, Bifurcations and Chaos in a Generic Management Model, **European Journal of Operational Research**, vol. 35, 1988, pp. 80-88.

Rasmussen, D.R., E. Mosekilde and J.D. Sterman, Bifurcations and Chaotic Behaviour in a Simple Model of the Economic Long Wave, **System Dynamics Review**, vol. 1, 1985, pp. 92-110.

Ravenstein, E.G., The Laws of Migration, **Journal of the Royal Statistical Society**, vol. 48, part 2, 1885, pp. 167-235.

Reggiani, A., **Spatial Interaction Models: New Directions**, Ph.D. Dissertation, Department of Economics, Free University, Amsterdam, 1990.

Reggiani, A. and S. Stefani, Aggregation in Decisionmaking: A Unifying Approach, **Environment and Planning A**, vol. 18, 1986, pp. 1115-1125.

Reggiani, A. and S. Stefani, A New Approach to Modal Split Analysis: Some Empirical Results, **Transportation Research B**, vol. 23, no. 1, 1989, pp. 75-82.

Reichlin, P., Equilibrium Cycles and Stabilization Policies in an Overlapping Generation Model with Production, **Journal of Economic Theory**, vol. 40, 1985, pp. 89-102.

Reilly, W.J., **The Law of Retail Gravitation**, Knickerbocker Press, New York, 1931.

Reiner, R., M. Munz, G. Haag and W. Weidlich, Chaotic Evolution of Migratory Systems, **Sistemi Urbani**, vol. 2/3, 1986, pp. 285-308.

Ricci, G., A Differential Game of Capitalism: A Simulations Approach, **Optimal Control Theory and Economic Analysis** (G. Feichtinger, ed.), Elsevier, Amsterdam, 1985, pp. 633-642.

Rietveld, P., **Multiple Objective Decision Methods in Regional Planning**, North-Holland, Amsterdam, 1980.

Rietveld, P., Infrastructure and Regional Development. A Survey of Multiregional Economic Models, **The Annals of Regional Science**, vol. 23, no. 4, 1989, pp. 255-274.

Rosser, J.B., **From Catastrophe to Chaos: A General Theory of Economic Discontinuities**, Kluwer, Dordrecht, 1991.

Rössler, O.E., An Equation for Continuous Chaos, **Physics Letters A**, vol. 57, 1976, pp. 397-398.

Rouwendal, J., **Choice and Allocation Models for the Housing Market**, Kluwer, Dordrecht, 1989.

Roy, J.R. and J. Brotchie, Some Supply and Demand Considerations in Urban Spatial Interaction Models, **Environment and Planning A**, vol. 16, 1984, pp. 1137-1148.

Roy, J.R. and P.L. Lesse, Entropy Models with Entropy Constraints on Aggregate Events, **Environment and Planning A**, vol. 17, 1985, pp. 1669-1674.

Ruelle, D. and F. Takens, On the Nature of Turbulence, **Communications in Mathematical Physics**, vol. 20, 1971, pp. 167-192.

Rushton, G., Analysis of Spatial Behaviour by Revealed Space Preference, **Annals of the Association of American Geographers**, vol. 59, 1969, pp. 391-400.

Saalem, M., Siddiqui, V.S. and V. Gupta, Young Predation and Time Delays, **Mathematical Modelling of Environmental and Ecological Systems** (Shukla, J.B., T.G. Hallam and V. Capasso, eds.), Elsevier, Amsterdam, 1987, pp. 179-191.

Saarinen, T.F., **Environmental Planning, Perception and Behavior**, Hougton Mifflin, Boston, 1976.

Sayers, C.L., Statistical Inference Based upon Non-Linear Science, **European Economic Review**, vol. 35, no. 2/3, 1991, pp. 306-312.

Schaffer, W.M. and M. Kot, Differential Systems in Ecology and Epidemiology, **Chaos** (A.V. Holden, ed.), Manchester University Press, Manchester, 1986, pp. 158-178.

Scheinkman, J.A. and B. LeBaron, Nonlinear Dynamics and Stock Returns, **Journal of Business**, 1989, pp. 311-337.

Schuster, H.G., **Deterministic Chaos**, VCH, Veinheim, 1988.

Scott, A.J., A Model of Nodal Entropy in a Transportation Network with Congestion Costs, **Transportation Science**, vol. 5, 1971, pp. 204-211.

Scott, C.H. and T.R. Jefferson, Entropy Maximising Models of Residential Location via Geometric Programming, **Geographical Analysis**, vol. 9, no. 2, 1977, pp. 181-187.

Semple, R.K. and H.L. Gauthier, Spatial-Temporal Trends in Income Inequalities in Brazil, **Geographical Analysis**, vol. 4, 1972, pp. 169-180.

Sen, A., Modeling Spatial Flows, Unpublished Manuscript, School of Urban Planning and Policy, University of Illinois at Chicago, Chicago, Ill., 1982.

Sen A., Research Suggestions on Spatial Interaction Models, **Transportation Research A**, vol. 19, no. 5/6, 1985, pp 432-435

Sen, A. and Pruthi, R.K., Least Squares Calibration of the Gravity Model When Intrazonal Flows Are Unknown, **Environment and Planning A**, vol. 15, 1983, pp. 1545-1550.

Sen, A. and S. Soot, Selected Procedures for Calibrating the Generalized Gravity Model, **Papers of the Regional Science Association**, vol. 48, 1981, pp. 165-176.

Senior, M.L. and A.G. Wilson, Exploration and Synthesis of Linear Programming and Spatial Interaction Models of Residential Location, **Geographical Analysis**, vol. 6, 1974, pp. 209-238.

Serra, R. and M. Andretta, **Introduzione alla Fisica dei Sistemi Complessi**, C.L.U.E.B., Bologna, 1984.

Shannon, C.W. and W. Weaver, **The Mathematical Theory of Communication**, University of Illinois Press, Urbana, Ill., 1949.

Sheffi, R.H. and C.F. Daganzo, On the Estimation of the Multinomial Probit Model, **Transportation Research Record A**, vol. 16, 1982, pp. 447-456.

Sheppard, E.S., Entropy in Geography: An Information Theoretic Approach to Bayesian Inference and Spatial Analysis, University of Toronto, Department of Geography, Toronto, Discussion Paper, no. 18, 1975.

Sikdar, P.K. and Karmeshu, On Population Growth of Cities in a Region: A Stochastic Nonlinear Model, **Environment and Planning A**, vol. 14, 1982, pp. 585-590.

Slater, P.B., Minimum Cross-Entropy and Spatial Interaction Dynamics, **Journal of Non-Equilibrium Thermodynamics**, vol. 13, no. 4, 1988, pp. 399-402.

Smith, T.E., On the Relative Frequency Interpretation of Finite Maximum-Entropy Distributions, Regional Science Research Institute, Philadelphia, 1972.

Smith, T.E., A Cost Efficiency Principle of Spatial Interaction Behavior, **Regional Science and Urban Economics**, vol. 8, 1978a, pp. 313-337.

Smith, T.E., A General Efficiency Principle of Spatial Interaction, **Spatial Interaction Theory and Planning Models** (Karlqvist, A., L. Lundqvist, F. Snickars and J.W. Weibull, eds.), North Holland, Amsterdam, 1978b.

Smith, T.R. and W.A.V. Clark, Housing Market Search Behavior and Expected Utility Theory: I. Measuring Preferences for Housing, **Environment and Planning A**, vol. 14, 1982, pp. 681-698.

Snickars, F. and J.W. Weibull, A Minimum Information Principle, **Regional Science and Urban Economics**, vol. 7, no. 1/2, 1977, pp. 137-168.

Somermeijer, W.H., **Specificatie van Economische Relaties**, Reprint no. 126, Econometrisch Instituut, Erasmus University, Rotterdam, 1966.

Somermeijer, W.H. and R. Bannink, **A Consumption-Savings Model and its Applications**, North-Holland, Amsterdam, 1973.

Sonis, M., Spatio-Temporal Spread of Competitive Innovations: an Ecological Approach, **Papers of the Regional Science Association**, vol. 52, 1983a, pp. 159-174.

Sonis, M., Competition and Environment: A Theory of Temporal Innovation Diffusion, **Evolving Geographical Structures** (Griffith, D.A. and A.C. Lea, eds.), Martinus Nijhoff, The Hague, 1983b, pp. 99-129.

Sonis, M., Dynamic Choice of Alternatives, Innovation Diffusion and Ecological Dynamics of Volterra-Lotka Model, **London Papers in Regional Science**, vol. 14, 1984, pp. 29-43.

Sonis, M., A Unified Theory of Innovation Diffusion, Dynamic Choice of Alternatives, Ecological Dynamics and Urban/Regional Growth and Decline, **Ricerche Economiche**, vol. XL, no. 4, 1986, pp. 696-723.

Sonis, M., Universality of Relative Socio-Spatial Dynamics, **Occasional Paper Series on Socio-Spatial Dynamics**, vol. 1, no. 3, 1990, pp. 179-204.

Sonis, M. and D.S. Dendrinos, Period-Doubling in Discrete Relative Spatial Dynamics and the Feigenbaum Sequence, **Mathematical Modelling: an International Journal**, vol. 9, 1987, pp. 539-546.

Sparrow, C.T., **The Lorenz Equations: Bifurcations, Chaos and Strange Attractors**, Springer, New York, 1982.

Sterman, J.D., A Behavioural Model of the Economic Long Wave, **Journal of Economic Behaviour and Organization**, vol. 5, 1985, pp. 17-53.

Sterman, J.D., Deterministic Chaos in Models of Human Behaviour: Methodological Issues and Experimental Results, **System Dynamics Review**, vol. 4, 1988, pp. 148-178.

Stetzer, F. and A.G. Phipps, Spatial Choice Theory and Spatial Difference: A Comment, **Geographical Analysis**, vol. 9, no. 4, 1977, pp. 401-403.

Stewart, I., **Does God Play Dice**?, Basil Blackwell, Oxford, 1989.

Stewart, J.Q., An Inverse Distance Variation for Certain Social Influences, **Science**, vol. 93, 1941, pp. 89-90.

Stopher, P.R., A.H. Meyburg and W. Brög (eds.), **New Horizons in Travel-Behavior Research**, Lexington Books, Lexington (Mass.), 1981.

Stouffer, S.A., Intervening Opportunities: A Theory Relating Mobility and Distance, **American Sociological Review**, no. 5, 1949, pp. 845-867.

Stouffer, S.A., Intervening Opportunities and Competing Migrants, **Journal of Regional Science**, vol. 2, 1969, pp. 1-26.

Sturis, J. and E. Mosekilde, Bifurcation Sequence in a Simple Model of Migratory Dynamics, **System Dynamic Review**, vol. 4, no. 1/2, 1988, pp. 208-216.

Stutzer, M., Chaotic Dynamics and Bifurcation in a Macro Model, **Journal of Economic Dynamics and Control**, vol. 2, 1980, pp. 253-275.

Taha, H.A., **Operations Research**, MacMillan, New York, 1986.

Tan, K.C. and R.J. Bennet, **Optimal Control of Spatial Systems**, George Allen & Unwin, London, 1984.

Theil, H., **Economics and Information Theory**, North Holland, Amsterdam, 1967.

Thom, R., **Structural Stability and Morphogenesis**, Addison-Wesley, Reading, 1975.

Timmermans, H., Decision Models for Predicting Preferences among Multiattribute Choice Alternatives, **Recent Developments in Spatial Data Analysis** (Bahrenberg, G., M.M. Fischer and P. Nijkamp, eds.), Gower, Aldershot, Hants, 1984, pp. 337-354.

Timmermans, H. and A. Borgers, Spatial Choice Models: Fundamentals, Trends and Prospects, Paper presented at the Fourth Colloquium on Theoretical and Quantitative Geography, Veldhoven, 1985.

Timmermans, H. and A. Borgers, Dynamic Models of Choice Behaviour: Some Fundamentals and Trends, **Urban Dynamics and Spatial Choice Behaviour** (Hauer, J., H. Timmermans and N. Wrigley, eds.), Kluwer Academic Publishers, Dordrecht, 1989, pp. 3-26.

Tinbergen, J., **Economic Policy: Principles and Design**, North-Holland Publ. Co., Amsterdam, 1956.

Tobler, W., An Alternative Formulation for Spatial Interaction Modelling, **Environment and Planning A**, vol. 15, 1983, pp. 693-703.

Tomlin, S.G., A Kinetic Theory of Urban Dynamics, **Environment and Planning A**, vol. 11, 1979, pp. 97-105.

Tuma, N.B. and M.T. Hannan, **Social Dynamics: Models and Methods**, Academic Press, New York, 1984.

Turner, J., Non-Equilibrium Thermodynamics, Dissipative Structures and Self-Organization, **Dissipative Structures and Spatio-Temporal Organization Studies in Biomedical Research** (Scott, G. and J. MacMillan, eds.), Iowa State University Press, Iowa, 1980, pp. 13-52.

Velupillai, K., A Neo-Cambridge Model of Income Distribution and Unemployment, **Journal of Post-Keynesian Economics**, vol. 5, 1983, pp. 454-473.

Vicsek, T., **Fractal Growth Phenomena**, World Scientific, Singapore, 1989.

Vollering, J.M.C., **Care Service for the Elderly in the Netherlands**, Ph.D. Dissertation, Tinbergen Institute Research Series, no. 13, Amsterdam, 1991.

Volterra, V., Variazione e Fluttuazione del Numero di Individui in Specie Animali Conviventi, **Mem. Accad. Nazionale Lincei**, vol. 31, ser. 6, no. 2, 1926, pp. 31-113.

Vorst, A.C.F., A Stochastic Version of the Urban Retail Model, **Environment and Planning A**, vol. 17, 1985, pp. 1569-1580.

Webber, M.J., Entropy-Maximising Models for the Distribution of 'Expenditures', **Papers of the Regional Science Association**, vol. 37, 1976, pp. 185-200.

Wegener, M., F. Gnad and M. Vannahme, The Time Scale of Urban Change, **Advances in Urban System Modeling** (Hutchinson, B. and M. Batty, eds.), North-Holland, Amsterdam, 1985, pp. 175-197.

Weibull, J.W., An Axiomatic Approach to the Measurement of Accessibility, **Journal of Regional Science and Urban Economics**, vol. 6, 1976, pp. 357-379.

Weibull, J.W., A Search Model for Microeconomic Analysis with Spatial Applications, **Spatial Interaction Theory and Planning Models** (Karlqvist, A., L. Lundqvist, F. Snickars and J.W. Weibull, eds.), North Holland, Amsterdam, 1978, pp. 47-53.

Weidlich, W. and G. Haag, **Concepts and Models of a Quantitative Sociology**, Springer, Berlin, 1983.

White, R.W., Transitions to Chaos with Increasing System Complexity: The Case of Regional Industrial Systems, **Environment and Planning A**, vol. 17, 1985, pp. 387-396.

Williams, H.C.W.L., Travel Demand Models, Duality Relations and User Benefit Relations, **Journal of Regional Science**, vol. 16, no. 2, 1976, pp. 147-166.

Williams, H.C.W.L. and A.G. Wilson, Some Comments on the Theoretical and Analytic Structure of Urban and Regional Models, **Sistemi Urbani**, vol. 2/3, 1980, pp. 203-242.

Wilson, A.G., A Statistical Theory of Spatial Distribution Models, **Transportation Research**, vol. 1, 1967, pp. 253-269.

Wilson, A.G., **Entropy in Urban and Regional Modelling**, Pion, London, 1970.

Wilson, A.G., **Catastrophe Theory and Bifurcation**, Croom Helm, London, 1981.

Wilson, A.G., Comments on Alonso Theory of Movement, **Environment and Planning A**, vol. 12, 1980, pp. 727-732.

Wilson, A.G. and R.J. Bennet, **Mathematical Methods in Human Geography and Planning**, J. Wiley & Sons, Chichester, New York, 1985.

Wilson, A.G. and M.J. Kirkby, **Mathematics for Geographers and Planners**, Clarendon Press, Oxford, 1980.

Wilson, A.G. and M.L. Senior, Some Relationships between Entropy Maximising Models, Mathematical Programming Models and Their Duals, **Journal of Regional Science**, vol. 14, no. 2, 1974, pp. 207-215.

Wilson, A.G., A.F. Hawkins, G.J. Hiu and D.J. Wagon, Calibration and Testing of the SELNEC Transport Model, **Regional Studies**, vol. 3, 1969, pp. 337-350.

Wissen, L.J.G. van and A. Rima, **Modelling Urban Housing Market Dynamics**, North-Holland, Amsterdam, 1988.

Wolf, A., J.B. Swift, H.L. Swinney and J.A. Vastano, Determining Lyapunov Exponents from a Time Series, **Physica D**, vol. 16, 1985, pp. 285-317.

Wrigley, N., Quantitative Methods: Developments in Discrete Choice Modelling, **Progress in Human Geography**, vol., 6, 1982, pp. 547-562.

Wrigley, N., Categorical Data Analysis and Discrete Choice Modeling in Spatial Analysis, **Measuring the Unmeasurable** (Nijkamp, P., H. Leitner and N. Wrigley, eds.) Martinus Nijhoff, Dordrecht, 1984, pp. 115-140.

Yamamura, E., Synergetic Location Theory, **Environmental Sciences**, Hokkaido University, vol. 10, no. 1, 1987, pp. 71-80.

Yorke, J.A. and E.D. Yorke, Chaotic Behaviour and Fluid Dynamics, **Hydrodynamics Instabilities and the Transition to Turbulence** (Swinney, H.L. and J.P. Gollub, eds.), Springer Verlag, Berlin, 1981, pp. 77-92.

Young, E.C., The Movement of Farm Population, Ithaca, N.Y., **Cornell Agricultural Experiment Station Bulletin**, no. 426, 1924.

Zhang, W.B., **Economic Dynamics**, Springer-Verlag, Berlin, 1990.

Zhang, W.B., **Synergetic Economics**, Springer-Verlag, Berlin, 1991.

Zipf, G.K., **Human Behaviour and the Principle of Least Effort**, Cambridge, Addison-Wesley Press, 1949.

Zipser, T. (ed.), The Model of Intervening Opportunities in Theory and Practice of Territorial Arrangement, **Scientific Papers of the Institute of History of Architecture, Arts and Technology of the Technical University of Wroclaw**, no. 21, Wydawnictwo Politechniki Wroclawskiej, Wroclaw, 1990.

Zuylen, H.J. van, The Information Minimising Method: Validity and Applicability to Transportation Planning, **New Developments in Modelling Travel Demand and Urban Systems** (Jansen, G.R.M., P.H.L. Bovy, J.P.J.M. van Est and F. Le Clercq, eds.), Saxon House, Farnborough, 1979, pp. 344-371.

Index

accessibility 7, 15, 29, 30, 41, 69, 94, 103, 107, 109, 116, 150, 213, 221

additivity conditions 5, 6, 9, 19, 24, 27, 35, 37, 56, 174

aggregate models 60, 71

aggregate utility theory 13

Alonso model 28, 29, 69, 71, 72, 233, 243

attraction-constrained SIM 8

attractor 122, 123, 124, 128, 129, 131, 132, 136, 155, 156, 157, 158, 159, 182

balancing factors 6, 7, 9, 22, 26, 28, 29, 31, 111

Bayesian statistics 35, 90

bifurcation 60, 89, 91, 120, 121, 123, 124, 127, 128, 137, 142, 143, 146, 147, 150, 159, 170, 182, 183, 200, 202, 209, 216, 221, 231, 232

bifurcation diagram 182, 183, 184, 186, 188

Brownian motion 96, 99, 104

capture probability 102

catastrophe behaviour 90, 91, 102, 103, 104, 106, 107, 108, 116 120, 234, 239

chaos theory, 89, 95, 96, 119, 125, 133, 151, 165, 166, 233, 235, 243

chaotic behaviour 124, 125, 126, 134, 135, 139, 140, 143, 147, 161, 166, 170, 176, 177, 179, 182, 190, 191, 209, 213, 216, 217, 222, 234, 235, 237, 247, 240

chaotic motion 123, 127, 130, 140, 143, 146, 150, 156

closed orbit 122, 202, 216

competing destination models 7, 69

competition coefficient 213, 215, 216, 222

competition model 200, 210, 235

complex system 126, 133, 166, 170, 182, 190, 192

congestion effect 72, 73, 74, 75, 76, 83, 85, 90, 92, 107

cost efficiency 12

cost friction coefficient 40, 46, 56

decompositional multi-attribute preference model 66

delay effects 180, 181, 190, 192

deterministic system 135

deterministic utility theory 62, 63

deterrence function 7, 11, 23

diffusion component 99, 101, 104, 119

disaggregate models 60, 61, 62

distance friction 5, 7, 19, 23, 24, 106

dogit model 66

doubly-constrained SIM 28, 29, 41, 69, 71, 82, 108, 233

dynamic logit models 165, 166, 173, 180, 192

economic utility approach 30

elimination-by-aspects models 65

entropy 3, 16, 17, 18, 19, 20, 21, 22, 24, 25, 26, 27, 28, 29, 30, 31, 32, 34, 35, 36, 37, 38, 39-42, 44-48, 50, 52-56, 59, 60, 64, 68, 70, 71, 82, 83, 89, 90, 91, 93, 94, 95, 100, 102, 105, 107, 108, 109, 112, 132, 166, 233, 234, 243

equilibrium point 23, 108, 117, 121, 152, 202, 211

escape frequency 101

extreme value models 65, 66, 71

Feigenbaum route 128

fixed point 123, 161, 162, 215, 230, 231

fractal 133, 139, 155, 243

geometric programming 5, 39, 42, 43, 44, 46, 47, 53, 54, 55

gravity theory 3, 4, 5, 7, 8, 10, 11, 13, 16, 60, 70, 233

Hamiltonian 109

Hopf bifurcation 123, 124, 142, 143, 150, 209, 213, 216, 221, 222, 231, 232

IIA property 66, 69, 71, 82, 233

information theory 18, 26, 34, 132

intermittency 177, 178, 179

intervening opportunities model 5

kinetic theory 102

Kolmogorov entropy 132

Lagrangian 21, 109, 114

Liapunov exponent 130, 131, 132, 134

limit cycle 122, 123, 129, 131, 213

linear programming 3, 39, 40, 48, 51, 52, 54

logistic equation 91, 127, 128, 150, 168, 174, 179, 180, 192, 211

Lowry model 16

marginal utility function 166, 235

master equation 97, 139, 240

maximum likelihood approach 37

multinominal logit model 65, 69, 70, 71, 72, 101, 103, 107

multinomial probit model 65, 66

nested logit model 65, 66, 69, 70, 71, 82, 97

Newton's law 3, 4

non-linear dynamics 120, 121, 124, 126, 210

onset of instability 139, 142, 143, 146, 174, 190

optimal control entropy model 93, 108, 109, 112

oscillating behaviour 143, 149, 172, 173, 174, 200, 202, 205, 216, 221

periodic solution 122

Poincaré-Bendixson theorem 123, 156

prey-predator system 121, 166, 168, 170, 206, 210

prior information 28, 34

probabilistic utility approach 31

production-constrained SIM 8, 29, 69, 102, 103, 104

prominence theory of choice 65

random noise 95, 97

random utility theory 59, 60, 62, 63, 69, 70, 82, 166

Ruelle-Takens-Newhouse route to chaos 129

sequential choice 67, 71, 82, 233

stochastic calculus 99

stochastic differential equation 99, 104

stochastic dominance 68

stochastic optimal control 89, 97, 100, 102, 104, 108, 113, 116

stochastic process 96, 98, 99, 101, 104

strange attractor 123, 124, 127, 128, 129, 132, 139, 155, 158, 159, 182, 192, 235

structural (in)stability 107, 121, 234

synergetic models 203, 209, 222, 235

systemic variables 28, 29

trip assignment model 40, 41, 54

unconstrained SIM 8, 44

welfare function 13, 150

white noise 98, 99, 101, 239